Ecological Readings
in
The Veda

Reconstructing Indian History & Culture
(ISSN 0971-3824)

1. Society in Ancient India; by Sures Chandra Banerji
2. Political Thought in Ancient India; by G.P. Singh
3. Early Indian Historical Tradition and Archaeology; by G.P. Singh
4. Vasantotsava: The Spring Festivals of India; by Leona M. Anderson
6. Vṛndāvana in Vaiṣṇava Literature; by Maura Corcoran
7. Social History of the Tamils (1707-1947); by P. Subramanian
8. Ancient Indian Coinage; by Rekha Jain
9. Buddhism and Socio-Economic Life of Eastern India; by Bimal Chandra Mohapatra
10. Chandragupta Maurya; by P.L. Bhargava
11. Mṛtyu: Concept of Death in Indian Traditions; by Gian Giuseppe Filippi
12. Yama: The Glorious Lord of the Other World; by Kusum P. Merh
13. Sacred Complex of Ujjain; by D.K. Samanta
14. Vedic View of the Earth; by S.R.N. Murthy
15. Society in the Atharvaveda; by B.S. Kharade
16. Orissan History, Culture and Archaeology; by S. Pradhan
17. Education in Ancient India; by Mitali Chatterjee
18. Surā: The Liquor and the Vedic Sacrifice; by M.B. Kolhatkar
19. Human Ecology in the Vedas; by Marta Vannucci
20. Governance in Ancient India; by Anupa Chandra Pandey
21. The Cultural Glory of Ancient India: A Literary Overview; by Sures Chandra Banerji
22. Kauṭilya's Arthaśāstra in the Light of Modern Science & Technology; by Sunil Sen Sarma
23. History of Jainism : With Special Reference to Mathurā; by V.K. Sharma
24. Cultural Tourism in India: Museums, Monuments & Arts — Theory and Practice; by S.P. Gupta; Krishna Lal; Mahua Bhattacharya
25. Ancient Indian Historiography : Sources and Interpretations; by G.P. Singh
26. Republics, Kingdoms, Towns and Cities in Ancient India; by G.P. Singh.
27. Facets of Ancient Indian History and Culture: New Perception; by G.P. Singh
28. Kamakhya : A Socio-cultural Study; by Nihar Ranjan Mishra
29. History and Culture of Tamil Nadu as Gleaned from the Sanskrit Inscriptions Vol. 1 (up to *c.* AD 1310); by Chithra Madhavan
30. Lifestyle of the Vedic People; by Pranati Ghosal
31. History and Culture of Tamil Nadu as Gleaned from the Sanskrit Inscriptions Vol. 2 (up to *c.* AD 1310 - *c.* AD 1885); by Chithra Madhavan
32. Matrimonial Alliances and Ancient Indian Polity (*c.* 600 BCE to *c.* 650); by Preeti Prabhat
33. Artisans and Craftsmen of Northern India; by Kuldeep Singh Thind
34. The Evolution of Historiographical Tradition in Ancient and Medieval India; by G.P. Singh
35. Warfare in Ancient India : Theory and Practice; by Soma Basu
36. Sanskrit Education and Literature in Ancient and Medieval Tamil Nadu; by Chithra Madhavan

Reconstructing Indian History & Culture, no. 5

Ecological Readings in The Veda

Matter - Energy - Life

M. Vannucci

DK PRINTWORLD
Publishers of Indian Traditions

Cataloging in Publication Data — DK

[Courtesy: D.K. Agencies (P) Ltd. <docinfo@dkagencies.com>]

Vannucci, M. (Marta), 1921-

Ecological Readings in the Veda.

(Reconstructing Indian history & culture ; no. 5)

Includes bibliographical references (p.

Includes index.

ISBN: 9788124600092

1. Ecology in the Vedas. 2. Vedas — Criticism, interpretation, etc. I. Title. II. Series: Reconstructing Indian history & culture ; no. 5.

ISBN: 978-81-246-0009-2

First published in India in 1994

Third impression in 2016

Printed and published by:

D.K. Printworld (P) Ltd.

Regd. office : *Vedaśrī*, F-395, Sudarshan Park

Ramesh Nagar Metro Station

New Delhi – 110 015

Phones : (011) 2545 3975; 2546 6019; *Fax* : (011) 2546 5926

E-mail: indology@dkprintworld.com

Web: www.dkprintworld.com

To

Erico *and* **Dino**

to whose incessant questioning
from the days of their first
'how' and 'why', 'yes' and 'no'
I owe much of my understanding

Yes, at thy birth, when thou hadst sat in glory
    thou, Agni, wast the aim of invocations.
The Gods came near, obedient to thy summons
    and thus obtained their ranks as chief Protectors.

Ṛ̣V. X; 6, 7. (transl. Griffith)

Foreword

Dr. Marta Vannucci, a Biologist, who has spent many decades studying mangroves tropical forests and the ecology of this earth has turned her attention to ecology of matter, energy and life as reflected and articulated in the *Vedic* corpus. She is neither scientist turned meta-physician nor agnostic turned believer, nor a materialist turned spiritualist. She is a scientist, exploring, investigating through textual sources what she has known and experienced as an investigative scientist through empirical studies.

This distinguishes the present work and the author's quest, from many others who have either tried to justify ancient knowledge on the basis of modern scientific theories or those others who have declared that all was known and there was nothing more to know.

Dr. Marta Vannucci clearly states the problem as also the hypothesis which she has set herself. As a biologist, she attempts to decipher the code with a biological key of the textual tradition. Earlier, people drawn from other disciplines, especially from linguistics, mathematics, physics and astronomy have tried to interpret the *Vedas* from the point of view of their particular discipline. Marta Vannucci undoubtedly uses the biological sciences for interpreting the text but focusses attention on aspects of matter and energy as personified by the *Vedic* God easily recognised as Agni. The relationship of Agni with Savitṛ symbolising life and even hope are her pointed concerns.

The *Ṛgveda* begins with that memorable verse - We worship Agni, the *Purohit*, God, minister of sacrifice . . . and throughout this *Veda* as also the *Sāmaveda* and *Atharvaveda*, Agni is primary concern of man as a species. Dr. Vannucci incisively analyses, the approach to the principle of energy and fire contained in the *Vedic* corpus and relates it to the evidence of similar preoccupation of early man in sources as far apart as those of Stonehenge and the Greek myths of Prometheus.

The implications of man's preoccupation with fire both as the principle of light in the sun and the control of domestic fire on earth and therefore the recognition of fire as the primordial element for maintaining cosmic order (Ṛta) are immense. This is where the conception of Agni and the notion of ecological balances become relevant. Neither less or more, neither uncontrolled or over-controlled fire is indispensable for creativity with potent power of creation and destruction. Man's intelligence alone can be controller and wielder.

Vannucci dwells on these aspects with great convictions in her chapters dealing with ecological aspects and the relationship of Lord Agni and Fire. From this she proceeds to make the relationship between fire and the Lord Savitṛ. Now she moves on how Savitr is on the one hand an aspect of Varuṇa, i.e., water and on the other, life itself. After persuing interpretations of Lord Savitṛ, Dr. Vannucci identifies Savitṛ as the essence of life itself and according to the biologist, she reminds us that Savitṛ is the first and the foremost Vivifier, the one which gives life because it is he who is the life principle. Savitṛ is therefore the source of life, the same life that he ceaselessly regulates. The relationship of Agni and Savitṛ is at multiple levels of fire and water, of energy and life and it is pertinent to remember that unlike other personifications Savitr is without gender.

Vannucci concludes by drawing all strands together of the textual sources and then relates them to the contemporary discourse of ecology. This comparatively young discipline points at the inter relationships of organic matter and energy of living beings among themselves and with the environment. The environment in turn comprises both organic and non-organic matter, i.e., both living and non-living. The *Vedic* man's pre-occupation by whatever route he came to it, reflects as deep a concern with the nature of matter, organic or inorganic and the nature of inter relationships of all living beings, human, animal, plant, i.e., all biological species. Obviously in the *Ṛgveda*, man did not postulate a scientific theory but it is equally clear that he was aware and self conscious of the principles on which scientific theories are based and more he was acutely perceiving of the phenomenon of life. He reflected on the question what was the relationship between him and the life around he saw, observed, experienced, without totally understanding the dynamic of structures of each species of organic matter. It would appear from the decoding done by Dr. Vannucci that the *Vedic* man through observation and reflection and not through laboratory experimentation or collection of data and then deducing theories was aware of the dynamic processes at work in this universe. This enabled him to identify the principle of the constant changes which were taking place between matter and energy and to explicate the order of the universe called Ṛta. This articulation is as impressive as it is remarkably valid today.

I hope that Biologists and Indologists will both read this book. It will stimulate thinking and provoke questions. Dr. Vannucci has triggered a line of thought which I hope many others will pursue.

I am both privileged and honoured that she should have asked me to write the Foreword. Inadequate as I may feel, perhaps one ecological principle is that those who know and those who want to know balance each other.

Dr. Kapila Vatsyayan

Preface

Ethics is the Soul of Biology
while
Biology is the body of Ethics.
M. Vannucci

The double and even triple meaning of the words and sentences in the *Veda* is a well known fact. To extricate some hidden or secondary meaning from the *mantras* I have used a biased approach of studying them from a biological perspective. Throughout the *Śāstra* one can sense the passionate urge of man for understanding-this is the basis of all science. Interspersed in the texts are hints of the application to practical purposes of the knowledge acquired, these are technicalities or technologies. Once the ideas were conceived, the concepts formulated and principles enunciated, the understanding was crystallized by deifying the same principles or **truths** perceived by man. Honoured and successful leaders, chiefs, kings, conquerors were also deified in the course of time. Myths were created to transmit the **truths** to the masses in an interesting and easily memorizable manner and to **explain** what was not yet really and fully understood. In all cultures myths and legends were also created to engrave in the collective memory the traditions and history of a people, of a clan or a race. Legends are usually the poetical presentation of history or are just wishful thinking.

In every society a group of persons stands aside as the learned men, the intellectuals and scientists who are variously called priests, sacerdotes, shamans, magicians, elders, *brāhmaṇas, paṇḍits,* scientists or philosophers. Thus two currents of culture are developed parallel to one another, the priestly and the popular. The knowledge and wisdom of the *Ṛṣi* grew alongside the culture of the other members of the Āryan clans. Also in pre-*Vedic* India a popular culture co-existed with the philosophical currents and later *Vedic* and pre-*Vedic* cultures co-existed side by side, influenced each other and were eventually fused in what is known as Hinduism. Due to an admirable attitude of mutual respect and tolerance the two, rather several traditions could survive almost intact, though variously modified through a continuous and active process of amalgamation, syncretism and evolution.

In the author's opinion Agni, (Fire) was the originally observed and used matter-energy system that became the primary primordial God, the *Ur-Gott.* Ṛta (Order) was the observed order and law that governs the physical universe or cosmos. 'Vāk', (Word or Speech), was the means of education and transmission of knowledge acquired through the grace of Agni. In conclusion the basis of the whole *Vedic* structure was philosophically scientific. Agni is male because of his strength and energy and Vāk is female because she generates and transmits knowledge and education. Ṛta is neutral because it is what **is**.

With these hypotheses on the meaning of Agni, Ṛta and Vāk and on the origin of philosophy, science and religion in mind, I sought corroboration in the ancient sacred texts and have compared different interpretations and translations, as they have come down to us. I hope eventually to elaborate a plausible theory.

I have reached to the conclusions that:

1. Because of man's urge, his need to understand and to know, philosophy in the original sense of the word, came

before religion. It came to man, by man, as science - both empirical and experimental.

2. Philosophy or science was translated into religion with its rites and rituals, experimental science was translated into technical practices or technology for the benefit of all.

3. Esoteric philosophies of an abstract type were later developed from religions and remained the domain of the initiated few. Some philosophies blend theory and practice, as for instance, different forms of *yoga*, which is a philosophical and biological discipline the ultimate goal of which is the fusion of the Individual Soul with the Cosmic Soul as it is based on discipline and control of the mind, body organs and systems. It is not a religion in the sense that it puts stress on discipline rather than on faith.

I believe the above to be the normal evolution of the human mind over successive generations, consequently similar sequences would have occurred in all cultures and would have followed comparable steps, such as:

1. Observation and logical deduction;
2. Empiricism and experimentation;
3. Development of a body of knowledge and formulation of prescriptions, not necessarily the formulation of generally valid theories;
4. Hermeneutics, or restriction of science and knowledge to the priestly classes, which includes scientists and technologists;
5. Hermeneutics becomes esoteric, limited to religious practices and tends to become stagnant; rigid rites become an end in themselves as recognized and prescribed physical and spiritual entelechies.

Ancient languages are more rigid and complex than modern simplifications, except for scientific and technical specialized jargons. Languages evolve continuously, e.g. Sanskrit into *Prākṛt* and Latin into neo-Latin languages. Simplification arises when root languages escape from the monopoly of the priesthood and the learned language gradually becomes amalgamated with the common people's language. Later on, secondary esoteric languages such as *tantra*, tibetan lamaism, modern sciences came into being; they are usually the result of cultural cross fertilization.

A more detailed study of Agni and Lord Savitṛ is given as well as a study of some ecological aspects of the *Veda*, mainly drawn from the *ṚV*, the *AV* and the *SV*.

I plead guilty for presenting a work that might seem, and to some extent rightly so, to have been published prematurily. Others, better qualified may wish to carry on, if this beginning is a worthwhile exercise as I think it is.

I am deeply indebted to our ancients to whom I offer deeply felt *praṇāma;* to Prof. R.N. Dandekar whose writings and living words marked the crystallization of many of my thoughts. Dr. Kapila Vatsyayan gave me unfailing encouragement and food for thought. I am specially grateful to Prof. B.N. Saraswati who gave me invaluable support through constant encouragement and advice over a span of few years. Long sleepless nights spent in quiet darkness and motionless contemplation, occasionally of distant and moonlit horizons in the Kumaon hills, I dare consider as the *primum mobile* of inspired insights. The constructive argumentative attitude of my younger friends, first among them Mrs. Anita Thapan sharpened my thoughts, checked my sometimes hurried enthusiasm over a *pāda* or other and added to my education. I am grateful to Miss Gitanjli Pandit for her careful revision of the manuscript and to Mrs. Yvette Field for her useful advice.

Who am I to thank these and many others? Rather I

would like to apologize for misreadings, for errors or misconcepts in understanding and for the unclearness in rendering. I stand to be corrected and welcome criticism.

Paraphrasing Prof. R.N. Dandekar who dedicates one of his books "to such of my pupils as do not regret having been my pupils", I would dedicate these notes of an ageing scientist to "such of my masters as do not know having been my masters". Many of them are anonymous or people from remote Indian villages.

M. Vannucci

Contents

Abbreviations

ABORI	*Annals of the Bhandarkar Oriental Research Institute*
AD	*A Domine*, After Christ
AV	*Atharvaveda*
BC	Before Christ
BP	Before Present
Br	*Brāhmaṇas* texts of the *Veda*
Br. Up	*Bṛhadāraṇyaka-Upaniṣad*
IE	Indo- European
MBh	*Mahābhārata*
MM	Max Müller
P	*Purāṇa*
Rm	*Rāmāyaṇa*
RV	*Ṛgveda*
Sam	*Saṁhita*
Smr	*Smṛti*
SpBr	*Śatapatha Brāhmaṇa*
Sr	*Śruti*
SV	*Sāmaveda*
SS and SV	Swami Satya Prakash Saraswati and Satyakam Vidyalankar
YV	*Yajurveda*

Introduction

Who, Agni, is thy kin of men?
    Who is thy worthy worshipper?
On whom dependent?
    Who art thou?

RV. I; 75, 3 (transl. Griffith)

Agni made ready the roads
    travelled by the Gods.

AV. XI; 1, 36

BROUGHT up with the critical Florentine frame of mind, and trained in scientific research methodology, I was, throughout long years of observing, thinking and reading, possessed by a few persistent questions. The inbuilt critical attitude of the Florentines produced masterpieces of the arts, of silk and woollen clothes, of sculpture and architecture, historiography and political sciences as well as a near perfect agricultural system. Florentines conceived and structured fundamentals of the contemporary experimental scientific edifice mainly through Galileo, his school and his successors who, among other things, were convinced that biological evolution is a particular aspect of the continuous evolution and change of everything. This humble essay addresses one of my quests: what was the evolution that has made man, MAN? Or, in other terms, when did man become the intellectually developed being we know and are, rightly or wrongly, so proud of? What are the links in the long chain of events that made man a living being with the power of discrimination? Or, as stated: "May he, Varuṇa, distinguish wisdom and folly of man" (*ṚV.* VII; 49, 3; *AV.* I; 33, 2).

I am convinced that animals think, but perhaps what made man one of a different kind was that he learned to draw conclusions that in turn influenced his behaviour. The power of discrimination is probably exclusive to man as is evidenced by the fact that responses to comparable or similar stimuli elicit different responses from different persons and from members of different cultural backgrounds; they even elicit different responses from one and the same person, according to the mood the person is in. In contrast, the responses of animals are stereotypes of the species. From thinking and understanding, man then had to transmit to his peers the knowledge or science that he had acquired. This he learned to do not only by direct example, but also by oral teaching: Vāk.

The *Veda* are at present the only recorded remains of the thoughts of ancient man, even if not of the earliest man. They are the only collection of thoughts and ideas whose length is above the minimum critical level that enables us to formulate a fairly clear image of how man thought a few millennia ago. The frequent reference to "our forefathers, to "the ancients", and to "tradition" indicate the antiquity of the widsom recorded in the *Ṛgveda* (*ṚV*) the *Atharvaveda* (*AV*) and the *Sāmaveda* (*SV*); it also shows that the *Veda* incorporate a corpus of knowledge acquired even before the oral tradition which we now have in written form, was organized in a systematic matter as the *Saṁhitās* (*Saṁ*).

What follows is the result of my endeavours to penetrate the minds of the ancients from my reading of the *Veda* and to try to understand how man may have become MAN; again the urge to understand! As a biologist, my interpretation is biased by trying to decipher the code with a bio-ecological key. My admittedly biased interpretation of the texts is a valid self-imposed criticism to the approach I used and the conclusions I draw. Throughout I have tried to understand the meaning condensed in the verses, the wording of which is subject to the constraints of the poetic metre used to facilitate memorization of the text. Most *Vedic* scholars have been linguists, etymologists, philosophers, mathematicians, physicists, astronomers, religious people, all of these branches of knowledge in which I claim no proficiency. However, not many, if any biologist has tried to plough his way laboriously through the *Veda*. I am fully aware that my work scratches only the surface of a treasure-mine of information, it is not more than a drop in the ocean and the only reason why I have succumbed to the pressure from many quarters to put some of my thoughts on paper is that due to my advanced age I will never be able to finish this work, even though it might never be possible even in a full lifetime. The *Veda* are a treasure-mine of information and this first contribution focuses only on aspects of matter and energy as personified by Lord Agni and on life and hope as personified by Lord Savitṛ and his associate *Bhaga*. Much more could be said on these topics and more will be added also on space and time considered as astronomical, biological and ecological cycles.

1

The Problem and the Hypotheses

"We worship Agni, the Purohit, God,
minister of sacrifice, who works
through eternal laws, who feeds and
sustains all that is divine and
luminous". . .

ṚV. I; 1,1

"Be of easy access to us, as a father to his son
Agni be with us for our sake". ..

ṚV. I; 1,9

TWO of the questions that haunted me throughout life are: when and how did man, the anatomical Hominid called *Homo sapiens* or, if your wish, *Homo sapiens sapiens* learn to use reason to behave accordingly? What contributed to develop his power of reason and discrimination?

The man we are talking about here is the zoological species to which we belong and of which we are so proud. May be once again Linné selected the correct epithet when he gave us the specific name *sapiens*, but still this does not imply that man is generally wise. *Sapēre* in Latin means to try, to have the taste of, or to taste the essence of understanding that leads to or should lead to knowledge and wisdom. Knowledge of something that is different, new and previously unknown, something that is more than just *scīre* that also means to know, but somehow means crude knowledge without the taste of it, if I may say so; like a student who memorizes a subject to pass the examinations, he "knows" the subject, but he does not really know it, he has not yet grabbed mentally the gist of it. Maybe it is the difference between erudition and knowledge. *Homo sapiens* is the zoological species that has the taste for, the firsthand experience of and the will to acquire and eventually use the understanding and wisdom derived from the observation of his inner and outer worlds. The urge to understand and thereby acquire knowledge is a basic trait of the human *forma mentis.* It is reflected in the common roots of such words as *scire, sapientia, sacrum, sacrifacère, science, sacred, sachh.*

By whatever means one gains an understanding, one must be able to think and I do not subscribe to the school of thought (Piaget, 1952) according to which primitive man shares with infants and animals the inability to think. Based on scientific experimental observation I believe that **thinking** is not exclusive to man, hence it is not this particular ability that differentiates man from animals. Indeed experimental psychology teaches, and among the first W Köhler showed with his experiments with chimpanzees carried out in Teneriffe during the first World War, that animals think. Recently Mandler (1990) has also disproved Piaget. Köhler's experiments culminated in his *Gestalt Theorie* (Köhler, 1947) that well demonstrates how qualitative observation, action and qualitative experimentation are inevitably linked together in this order The opportune and accurate reaction follows the holistic perception of the "surrounding" or "outside" world. Behaviour by trial and error does not proceed along the same steps.

We may visualize early man, or perhaps even the individuals of the species that preceded *Homo sapiens*, going beyond the consciously and perhaps subconsciously or even unconsciously rationalized perception of his inner and outer worlds by thinking and finally recognizing the relationship of cause and effect of the phenomena of the physical world. This I would call the *intelligĕre* phase; from *intelligĕre* the word intellect is derived. *Intelligĕre* literally means to tie, to unite, to relate one thing to others. It is the *scīre* and *sapēre* or the **testing** and **tasting**, as well as the *intelligĕre* phases that must necessarily come after the initial conscious endeavour to understand that which was observed; full understanding comes much later, if ever. To understand is *capīre*, that means to "fill fully", "to measure the fullness or capacity".

It thus happened that towards the end of a lifetime of studies I realized that it was from the *Veda* that I acquired my present convictions.

When the intellect perceives the truth, the seer, the *ṛṣi*, exclaims with joy: "now thou art well my Lord and shall be really happy" (*ṚV.* VIII; 1, 34) and Agni (l.c.. 1, 33) is the great giver. The mortal, troubled by the need to understand, may then come to one of the seven pathways that the wise have fashioned (*ṚV.* X; 5, 6). The seers with closed eyes understand through the intellect (*ṚV.* VIII; 25, 9), "even when they close their lids. . .they perceive".

According to Sāyaṇa (SS. and SV., p. 4380), in *ṚV.* X; 9, 1 "great happiness" means "perfect knowledge" or the transcendental knowledge of *Brāhman;* we could perhaps add-also of factual knowledge. The "first seed of spirit" I take as the desire to understand and is what rose in the beginning (ṚV. X; 129, 4); the *ṛṣis* could understand well what was not, or does not exist in the phenomenal world, but could not understand fully what they saw or sensed through the sense organs and had to admit painfully "not this, not that".

Again eagerness to understand, to acquire knowledge, is clear in *ṚV.* X; 88, 18: "how many are the fires and suns in number? What is the number of the dawn and waters? Not jestingly I speak to you, o Fathers, Sages, I ask you this for information" (transl. Griffith). Incidentally the verb "to inform" and the noun "information" that means "to form internally" is the same as the mathe- matician Ernst Mach said: "to understand something is to represent it mentally" or "visualize". This type of quest of the ancients is sheer scientific enquiry and endeavour. Agni is the one who responds to the question and is "he who will announce the sacrifice" (*ṚV.* X; 88, 17). "All that existed then was void and formless, by the great power of warmth, Agni, was born that Unit" (*ṚV.* X; 129, 3). "The seven seers have given birth to Agni" (*AV.* XI; 1,3), or "is born of the mouth of *Puruṣa*" (*AV.* XIX; 6,7). We may in other words say that the concept was "seen" or "formulated" that gave birth to the first God, Agni.

The eagerness and need to understand is eloquently expressed in *ṚV.* I; 71 and *ṚV.* I; 142, 6 where Agni signifies **Truth** or **Knowledge** and as such is the expression rather than the manifestation of God; he is the "loved and loving one" who, together with the Angirasas made the way to reach heaven and who burst the fortress of ignorance. Agni is the "treasure-lord of treasures" (*ṚV.* I; 72,1), forever granting all immortal bounties. In *ṚV.* I; 105 questions such as: "may I know this?" are repeatedly addressed on different topics and the refrain says: "why this, may I know, O Earth and Heaven? Which is the Truth?". Similar expressions of anxiety are formulated over and over again, as in *ṚV.*I; 75, 76 105, where the refrain implores: "let me know!". *AV.* XI; 5 goes a long way in eulogizing the *brahmachārī* and showing respect for him for his perseverance and learning. After listening to both concrete and incorporeal entities it is said that he is entitled to be respected because (*AV.* XI; 5, 20) "the herbs, past and future, day and night, the forest trees, the year together with the seasons-are all born (this I read as 'born in the mind', or 'are understood and explained') of the *Vedic* student". Questions frequently relate to natural cycles (*ṚV.* I; 113, 10) hence Ṛta or cosmic order. The ancient *ṛṣis* were successful also in applying philosophical principles to satisfy the basic human needs of solidarity, generosity and transfer of material, intellectual and spiritual culture.

Rather than exclusive knowledge of God, the *ṛṣis* sought all knowledge, or knowledge itself, because knowledge of the universe becomes knowledge of God. Thus as a consequence of the urge to understand, the concept or the idea of God must have taken shape in the mind of early man, long before the *ṛṣis* elaborated on the concept. God is the entity that **is** everything and **is** therefore self-conscious and omniscient, while man, realizes from experience, as can easily be demonstrated, that he is not and will never reach the state of ominiscience. Thus man cannot be equated to God and in turn God cannot be fully understood by man. So man called "God" all that he could not fully represent mentally. At a later stage, and this is the *Vedic* stage, man personified Gods, Ideas, Concepts to make them more readily intelligible; man can identify himself easily with personified ideas, concepts and gods and thus he binds himself to abstract ideas and concepts.

Agni is invoked: "shine this blessing on us, may we attain to perfect understanding"(*ṚV.* VII; 3,10) and "Agni stay with us nor far away from the Gods" (*ṚV.* X; 11,9). The antiquity of knowledge is proclaimed (ṚV. X; 17, 8) as personified by Saraswatī "who comes with the Fathers and rejoices with them".

Observation of the phenomena of the world around us is essential to relate cause and effect and thereby acquire knowledge. In fact we have all noticed how man, other mammals and birds, occasionally stop, look around, perhaps sniff the air and the ground or slightly change their posture, then trot or fly away in a chosen direction. We have all seen animals take detours to reach their objectives if there are impediments across the straight line, a fact that surely indicates that an analysis of the topographic situation was processed in the brain and was followed by instructions given to the locomotory organs. However, when and where the relationship between cause and effect breaks down or is not perceived by man, a sense of the supernatural, **God**, takes over.

Primitive man, hundred of thousands of years ago, must have observed nature closely and drawn conclusions to guide his actions. Beyond satisfying the urge of hunger and reproduction, some individuals must have willed to act on their environment to improve what we now would call "the quality of life". It is the evolution of the genetic complex that gave man the will to do, that in turn engineered *Homo sapiens.* The basic difference between man and the other primates and animals in general is that once the evolution of intelligence was set in motion, it could not be stopped because intelligence breeds constant self-improvement. Other Hominid species may have merged with ours or disappeared, while animals remain as they are or as far as they have intellectually developed for ever after.

MacKie (1988, p. 211) comments on what may have been the motives that moved the "wise men" of prehistoric times to build the megalithic structures of the British Isles, many of which are now proved to be some sort of solar calendars. MacKie "presumes" that the structures were made "for ceremonial reasons" and recognizes that "more than one intellectual leap must be made" to use the structures as calenders. He argues that "some reason would be behind the "systematic attempt. . .to define the exact length of the year. . ". I believe that the main reason is, as mentioned above, the human need to know and test one's knowledge or theory. The ancient Druids of the British Isles, or whoever the "wise men" would have been, were scientists in their own right, moved by the same urge of scientific discovery as the genuine contemporary scientist. The main difference lies in the way they expressed themselves. The many megalithic structures of Brittany in France and of the British Islands were described by Thom, who studied in detail some 60 sites (Ferguson, 1988, p. 32); this studies indicate that great pain was taken to place and build the structures, transport and erect megaliths of up to 20 tonnes in weight. In the Kumaon hills, Bisht and I have found some pre-historic megalithic structures that are aligned and oriented. We believe that they were used for ritualistic purposes probably associated to astronomical observations for the determination of calender fixes.

It is not possible to prove whether the detailed study came first and the ritual associated with it followed, but it is hardly to be doubted that at first a rough observation of solar, star and lunar movements had to be made and that subsequently this would be verified and the observations refined by the use of instruments such as the megaliths themselves. Megaliths, menhirs, stone slabs and rock alignments of Europe and those of Kumaon merely express the same basic urge to find out, to discover, to understand, to know, to unveil Ṛta and to test the truth of one's discovery. In one word, "to remove the veil of ignorance"

In conclusion, at one stage of the evolutionary process, man observing, thinking and then drawing reasoned deductions, must have formulated the idea of God **because** of the much that he understood and of the much more that he could not understand. He then learned to put to good use the expertise acquired and in fact Agni the God and agni the common fire have the same name. In the course of time, proud of their achievements, some men in all cultures equated themselves with God: "if I, O Indra were the Lord of riches!". . .(*ṚV.* VII; 32, 18). Others identified themselves with Agni, as did *Vaiśvānara* and *Prometheus* or they deified other men and in doing so they again equated themselves with the Gods.

Man however realized very early that ignorance breeds fear and that truth or knowledge is freedom from fear (*ṚV* I; 93). Such a powerful *ṛṣi* as Bharadvāja (*ṚV.* VI; 6, 1-3) reminds man of the multiple forms of Agni and that even the most humble form of fire must be kindled with "newest worship" because that is only as far as man can go, kindling the manifestation of God, but not beyond. Agni God is the "treasure Lord" (*ṚV.*I; 72) who "forever grants all immortal bounties" and Agni bursts the fortress of ignorance (*ṚV.* I; 71, 2). In the recommendation for worship the importance of ritual is implied; it is spelt early (*ṚV.* I; 1, 9-10) and it continued to be recognized throughout the history of man as very important; but in modern times often religious rites and rituals are replaced by civil ceremonies. Antoine de Saint-Exupéry in the 'Little Prince' bitterly regrets that rites are being forgotten and people can no longer afford the time to "dress their heart". After being instituted, rites tend to grow in complexity and punctiliousness and when mistakes are made other complex expiatory rites must be performed to purify and redeem the offender. In the very last hymn of the last *maṇḍala* of the *ṚV* the desirability is expressed of establishing a religious organisation or a "church". *ṚV.* X; 191 is a hymn to Agni: "assemble. . .let your minds be all of one accord. The place is common, common the assembly, common the mind, so be their thoughts united. A common purpose do I lay before you and worship with your general oblation. . .One and the same be your resolve and be your minds of one accord. United be the thoughts of all that all may happily agree" (transl. Griffith).

The evolution of Man the Thinker is well represented by the very ancient God Savitṛ, the Great Vivifier, the Great Stimulator, the Enlivener, closely associated to Ṛta (Dandekar 1979a, p. 16). Savitṛ is the expression of the idea and concept of Life, to this matter I will return later.

Observation must have led the most intelligent among early men to make and use fire and at the same time they would have observed that fire is terrible, fearsome and dangerous, as well as very useful. In an earlier paper (Vannucci: Fire and marksmanship as transcultural values; in press), I made the point that the ability to make and control fire and the ability to take accurate aim are the basic properties that differentiate man from animals, but here I am only concerned with Fire, the primordial God, the "eye and guardian of Ṛta" (*RV.* X; 8, 5).

Fire, as the most important element in the daily life of man and the one to which man owes the development of many of his techniques, had to be respected and venerated. Fire is multipurpose and multipotential, it is he "who travels in various directions" (*RV.* I; 141, 7) and he is omnipresent as he pervades everything (*RV.* I; 31); "Agni is the desire of what is and is to be" (*AV.* VI, 36, 3).

Man the Observer must have rationalized the many uses of fire while he was watching the outside world and the heat within himself. Not only the flames of fire must have entered into the realm of his thoughts, but a large number of other notions as well. Notions however are useless unless they are systematized, methodically and consciously. In trying to do this, man cannot have failed to observe the world's dynamics and as the mathematician Ernst Mach said that "to understand something is to represent it mentally" man would have applied his knowledge methodically to develop his techniques. Much has been said and written on the perception by early man of the lunar and solar cycles, of the seasons and cyclic phenomena in general, but I would like to refer here to more subtle observations of everyday life and death, those that shape the mind. One of them is the role of fire and heat in the normal functioning of life, another is that order and cycles.

The rural, the mountain and the sea people and those of us who have had the good fortune to be exposed to and guided into observing nature, cannot help but see order everywhere; from the symmetry or orderly asymmetry of germinating and growing plants, of flowers and of the arrangements of leaves and buds, to the strict laws that regulate the structure and function of the human and animal bodies and the rhythmicity of all natural phenomena, except for episodic events, such as an earthquake. Early man perceived, sensed, got impressed and lived by and with the order in nature. When ploughing started to be practiced, the convenience of ploughing through straight regular lines and seeding accordingly must have been as obvious as the regular alternation of even and odd numbers. Sītā, the furrow (*ṚV.* IV; 57, 6) is a basic aspect of Mother Earth, the good provider. The myth of the birth and death of Rāma's Sītā is a magnificent representation of the concept of cycles in nature further related to the *yoni*, as a symbol of femininity and thus of fecundity, generosity and compassion (Vannucci, 1992, p.25).

Much later Lavoisier in XVIII century France, expressed the principle of Conservation of Matter: "In nature nothing is created, nothing is destroyed but everything is transformed". Earlier the *ṛṣis*, for instance, Bharadwāja had expressed the same in a figurative manner, as in *ṚV.* VI where Agni Vaiśvānara, as *Jātavedas* and *Mātariśvan* is said to pervade everything, that he needs air to live (combustion), returns regularly as observer of Ṛta and as light he makes life possible on earth. The notion of order in nature is intuitively acquired through daily life an experience most regrettably denied to the majority of today's modern life systems.

Dante Alighieri (Paradiso I, 103 *et seq*) explains:. . . "*Le cose tutte quante - hannordine fra loro; e questo ĕ forma-che l'Universo a Dio fa simigliante*" (All things have order among themselves; and this is "form"-that makes the Universe similar to God). Dante follows the philosophy discussed by Saint Thomas Aquinas in the "*Summa Theologica*" (P.1, XV, art. 1) which is the cornerstone of dualistic scholastic philosophy according to which the world is distinct from God who created it alike, but not identical to himself: "*forma ad similitudinem cuius mundus est factus*". Differently the monism that pervades the *Śāstra* and culminated in *Vedānta*, identifies the whole universe with the *Brāhman* or Self and therefore everything is sacred since it is part of and not distinct from the

Godhead. Both philosophies, however, see **order** in all "creation" **that** is what was "**created** by God" or in "all the universe, that **is**" the Godhead or *Brahman* itself. While scholasticism sees order as a similitude, Vedism sees Ṛta as **the** reality and therefore an expression or manifestation of the Godhead that man can apprehend and must accept.

Chaos is a word invented late in the history of the intellectual development of mankind and mathematicians are now finding order even in "chaotic states", though a very special type of order, a sort of deterministic randomness. Order is everywhere the rule and disorder the exception, as is inherent in the concepts expressed by these two words. Order comes before disorder in the mind of children and simple people as well as observers of nature. Before order set in there was nothing because there was nothing to be arranged in space and time. Chaos sets in when order and randomness breakdown, this presupposes that before the event, things were in order. Chaos is frightening because it causes and demonstrates the breakdown of the rule that makes everything possible, including life, the rule that gives a meaning to everything and its parts and therefore the concept of chaos cannot have existed in the mind of man **before** the concept of order was mentally formulated from the observation of nature. Chaos and disorder are universally confusing because they conceal order, the order that is necessary for logical thinking. Before anything existed there was void, emptiness (*ṚV.* X; 129), but: " who knows? (*ṚV.* X; 129, 6), " the gods are later" (l.c. 7) "he, the first origin. . . knows it, or perhaps he knows not".

Order- *Ṛta*- is a basic concept derived, like the concept of the "fireness in fire" from observation followed by logical thinking. Since man is not the artisan or maker of nature, he cannot be responsible for the origin of order. The concept of Ṛta arose as an inescapable necessity to represent mentally the realities of the physical world: it is based on the principle that order reigns supreme and is expressed by the statement of laws on which scientific theories were later based and further developed. Ṛta is therefore, like Agni who is the essence of fire and who strengthens the concept of Ṛta as law (*ṚV.* I, 44, 14), a knowledge that must have been formed very early in the mind of man though neither Ṛta nor Agni are readily seen as a material entity. "By Ṛta, Ṛtā is restrained" (*ṚV.* IV; 3,9).

As physical and mental evolution of the human species progressed, at some very remote time man would have developed the ability to articulate stereotyped sounds to produce intelligible speech, a fact very important for the transfer of knowledge and for expressing abstract thoughts. The ability to pronounce conventional sounds that carry a specific meaning is evidenced, in the palaeontological record, by the development of the glottis and other anatomical structures, though unfortunately the glottis is not preserved well in the fossil record of the evolution of man. Though it is impossible to ascertain when and exactly how it happened, there is no doubt that the development of the organs of speech and of the corresponding parts of the brain went hand in hand. As speech and brain developed, so also did the ability to communicate with other men and to teach the younger generation, not only by example and demonstration, but also by explanation. The time when use and codification of sounds made into words was regulated to convey understanding, must have been a very exciting period in the history of mankind. It is not astonishing that later, Pāṇinī who codified the means of expression in almost mathematical terms, is respectfully revered for setting the rules of classical Sanskrit. Classical Sanskrit is more precise and complex than modern simplifications, as is obvious from what has happened in its evolution into Prākṛts or contemporary Sanskrit-derived languages.

Thus the concept of Vāk must be as old, or almost as old as the concepts of Agni and Ṛta. In the original order,

I would guess that the succession was thus: first was formed the universal concept of Energy and Power, then the concept of Order and Law in Time and Space and of Cause and Effect and then, lastly the concept of precise and correct communication: **Agni, Ṛta, Vāk** "Bṛhaspati gave name to objects, sent Vāk out first" (*Ṛ*V. X; 71,1). Words are sounds with a meaning and Vāk is sent out to assist with the power of words.

These ideas, if I am correct, would have followed from pure philosophical attitudes and procedures derived from the need and urge that man, or at least most men have (*ṚV.* III; 31) to rationalize, understand and communicate their findings; this would have satisfied some of the human eagerness to participate in the universal cosmic drama (*ṚV.* I; 10,11). . . "and (Indra) cause the seer to win a thousand gifts". Contemporary science progresses essentially along the same conceptual pathways; it is a repetition of the basic and essential frame of mind of man, nowadays enriched by powerful instrumentation and cumulative experience; experimentation, knowledge and communication. Basically Leonardo da Vinci's aphorism: "There is no truth that has not passed through our senses" is still valid if together with Leonardo we take truth as "understanding" and if we admit as Indian philosophies does that the mind is one of the sense organs. Dandekar writes (1979a, p. 62): "in the *Ṛg-veda*. . .we come across frequent reference to Ṛta, the cosmic-magic law which governs every happening in this world. . .The *Vedic* Ṛta. . .in a sense. . .is the *Vedic* version of the Hindu doctrine of *Karma*". The laws of nature are unchangeable (*ṚV.* I; 73,2), as "the solar rays diligently bring daylight" (*ṚV.* I; 3,8) and it is these that man endeavours to understand; man endeavours to acquire truth through the senses and finally represent the law, Ṛta, mentally.

A later epithet of Vāk (Speech) is *Sasarparī* or "swiftly moving" (*ṚV.* III; 53, 15); evidently the desire of fast and ever faster communication is not new to this century!

According to Sāyaṇa, Vāk being intelligence, dispelled the "unintelligence of the *Kuśikas*". The learned inherited the duty to teach other men and "loosen the invoker's thought, Indu, as they clean thee" (*ṚV.* IX; 95, 5) while Varuṇa is "held aloft above the ocean (of *avidya*) by the ancient God Trita" (*l.c.*; 95, 4) and the "secret names are uttered" (*l.c.*; 95.2). In *ṚV.* X; 71, Vāk is actually almost identified with *Jñānam*, or knowledge, a noun that appears only once and late in the *ṚV.* Vāk is invoked as the first, as a divinity who is concealed in the wealth of all subsequent auspicious knowledge; she represents oral tradition and symbolic speech; she represents sound or music and her power of discrimination is eulogized, including the hidden or esoteric meaning of all speech. The initiated understand the subtle meanings that are beyond the grasp of the non-initiated and only those who know the hidden meaning can follow the rituals profitably, with understanding, and they are those who come in triumph, having acquired knowledge. Vāk is also the means to convey prayers (*ṚV.* X; 114, 8) "Vāk spreads forth as far as prayer goes". . . (*l.c.*, 9). . .and conveys knowledge: "What Sage has learned the metres application? Who has gained Vāk, the spirit's aim and object?". "Vāk bears Mitra and Varuṇa, Indra and Agni" . . . "by her own nature knew the supreme *brāhman*" (*AV.* IV; 30)

Finally Vāk is feminine because she gives life to ideas and thoughts (*ṚV.* X; 125) "hear. . .all the truth as I (Vāk) declare it" (*ṚV.* X; 125, 4). Hence the importance of the word and of the name.

Interrelation of all natural laws would have gradually been perceived. But since man could not then, as we cannot today, perceive or define all the relations of cause and effect, he had to resort to *asu* (Dandekar 1979a, p. 55) or magical power, to "explain" the unexplainable. Thus to this day we still observe auspicious and unauspicious numbers, dates, hours and many other "superstitions"

The need to understand for the sake of having peace of

mind is well expressed again and again, as in *AV.* X; 7,4- *ṚV.* I; 76- *ṚV.* I; 75 and it includes the urge to apply the knowledge gained to practical purposes. The need to communicate is implicit in the very nature and essence of the *Veda*, thus Vāk is frequently invoked though often indirectly so (*AV.* VIII; 3, 6) and is often referred to as "the three syllables of the sacrifice". Agni is also often referred to indirectly as for instance in *AV.* XIII; 2, 46 when he is "coming like a milch-cow". Furthermore the dual divinity Indraagni that occurs not infrequently, indicates the concept of material and intellectual or spiritual understanding bound together.

In summary I believe that there is ample reason to conclude that philosophy as the search for knowledge came to man before religion. Philosophy is the science of primitive man who "wishes to understand". Philosophy and science are two words that express a host of ideas and concepts that were differentiated only much later in the process of systematization of man's knowledge. An Eskimo man told Rasmussen (Radin 1941, p. 50) that "our fathers were instructed by their fathers, they preserved ancient habits and rules that are the product of the experience and science of countless generations." This is absolutely correct because to a large extent science and tradition are the cumulative experience of many generations selected for what is good, positive and useful and expurgated of what is bad or ineffective. Tradition forms an important part of the corpus of all religions.

I use the words "philosophy" and "religion" in their original sense of "wish to understand" or "friend of knowledge" and "uniting or binding man to God" respectively, because the concept of God or the idea of a supreme Being and Ruler of all the world could have been conceived and could have originated only *after* the order prevailing in the world, or Ṛta, was perceived, recognized and unconditionally accepted. Only then could conceptual ideas be deified with consequent personification.

Radin (*l.c.*, p. 8) refers to religion as a "sentiment and an ensemble of acts, customs, beliefs and concepts associated to the sentiment. . .a mechanism invented by man". Here Radin contradicts himself, because if religion is a **sentiment** it cannot, implicitly and by exclusion, be a **mechanism invented** by man. Other modern "definitions" of the term "religion" fall into comparable contradictions.

Of the three basic concepts, Agni, Ṛta and Vāk, Agni *Jātavedas,* the all knower, is obviously the natural Ur-Gott of man. Recognized as the original energy and power, he is adorable and frightening, protector of good and bad (*ṚV.* I; 31,6); he is four-eyed and sees in all directions. He enlightens and gives knowledge and above all he gives the power of discrimination because he is the great Purifier, purifier of matter—concrete and abstract.

Agni, the Purifier par excellence (*AV.* XII; 5) extracts the essence and turns matter into spirit: "O Agni who separatest what was mixed" (*ṚV.* V; 8,3). One is tempted to paraphrase this into: "Lord Agni, who turns the material into ideas, concepts and knowledge of the Abstract". *ṚV.* I; 72, 2 is a very important hymn that goes a long way to show the dual personality of Agni as God and Agni as common fire. He is here recognized as the highest God by the other Gods who had not "found him" until they understood his philosophical meaning and knowledge. Agni is the giver of spiritual and material food and Agni is also the preserver of mysteries and rites; Agni is therefore also the *Hotṛ* par excellence and is the one with awful power (*ṚV.* I; 127,11-*ṚV.* V;10-*ṚV* X;71). He is the sagest of the Sages (*ṚV.* V; 42, 3), the all pervading energy though not specifically mentioned by name in some of the hymns to *Viśvedevāh* (*ṚV.* V; 41, 42). Agni (*ṚV.* VI, 1,1) is the worker of marvels and the bestower of strength and wisdom. Thus the study of Lord Agni became the corner-stone problem to be tackled before any other.

Finally I would like to say that I regard religions as the simplification of philosophy and science for the use of the common man since it is not possible or desirable to make philosophers of every human being. Religion presupposes the acceptance of dogmas as undisputed truths by which individuals should abide without hesitation or discussion, but with Faith. Religions demand unwavering faith on the part of the believer. To enforce or make attractive the dogmas, teachings and practices of all religions, rites and rituals are developed and mythologies are created to illustrate what is obvious and to produce a story behind what is not clear. Mythologies enable the masses to identify themselves with and thereby feel closer to the God they worship. Mythologies also bring easily accessible religious principles to the doorstep of every mind, whether simple or evolved. Finally mythologies go hand in hand with faith and tend to explain the truths of factual life as recognised by the "intellectuals" or "scientists" but for which a logical or experimentally reproducible result or immediate explanation has not always been found. There is much magic and there are many myths also in twentieth century science and much blind faith in the acceptance of its infallibility.

In conclusion, after intensive reading and re-reading of the sacred texts, my hypotheses in formulating this work are:

1. The will to know and to understand natural phenomena are the first movement of the human mind;
2. The observed relations of cause and effect or science, that cannot be fully understood are deified and/or explained as "magic";
3. Religion is born to unite man to the entity beyond his power of understanding; religion begins at the ultimate boundary of philosophy, where and when logical explanations of natural phenomena cannot readily be found;

4. Rites and rituals are developed to approach, propitiate and worship the ideas/concepts/ Gods;
5. Religions, with time, develop their own philosophies that are no longer empirical or experimental science.

The study of the *Veda* helped me to formulate these working hypotheses and guided me towards an understanding of the evolution of the intellect, science and knowledge of man.

"To me who understand hath Varuṇa spoken, the names borne by the Cow (Vāk) are three times seven;
The wise God, knowing the place's secret, shall speak to teach the race that cometh".

ṚV. VII; 87.4 (from Wilson and SS & SV transl.)

2

The Quest For Sources

"Agni the undeceivable, who spread out all the worlds,
keeper is he and guard of immortality"

ṚV. VI; 7, 7

THE fossil record bears witness to the evolution of man's body; the artifacts found in association with human fossil remains bear testimony to man's increasing technical performance; paintings and sculptures of animals, human figurines and "hominoids" as well as ornaments suggest the development and evolution of growing mental faculties. Much of what we know or presume we know at present about the evolution of man's mind is to a great extent the outcome of circumstantial rather than direct evidence. The ancient roots of man's way of thinking must be sought in the oral tradition; the folklore, the myths, the legends and the shamanistic practices and chants of the people of the world, while the ancient most are lost in past aeons. The *Vedic* lore is unique in that it has come down to us through an uninterrupted line of teachers with its roots and links in the most remote past. It is, however, a difficult task to find out what were the early thoughts and reasonings of man since the oral and written records begin at a very late stage of the evolution of man's mind; inevitably research in this field cannot but be speculative to some extent.

The *Veda* state the Truth — or Knowledge — but do not explain how this was acquired. For instance a hymn to

Dadhikrās (*RV.* VII, 44) appears to indicate the annual cycle of the sun and how to forecast the **calendar**, but no details are given on how the science was developed. The oral and practical instructions would be sufficient to transmit the expertise; the hymns of the *Veda* are often mainly mnemonic aids that express concepts and functions as represented by the Gods involved in that particular branch of science.

The evolutionary process of man's mind and the increasing development of manual ability took something like a million years or perhaps more to concretize, as indicated by palaentological remains of man-made fire. The oldest known remains of fire used by man for roasting purposes and/or sharpening tools comes from the Swartkrans cave in South Africa, and are dated approximately one to one and a half million years BP

Evolution was slow and our knowledge of the prehistory of man is fractional and to some extent conjectural. Early history is also insufficiently documented. Though the records of historical Egypt go back some millennia; much of the history of the development of thought and science of ancient Egyptians is lost. Chinese records also go back a few millennia; they have been rendered in English language and were studied in detail by the lifetime monumental work of Sir Joseph Needham, a distinguished biologist whose first book, published at a very eary age, is a landmark treatise on Experimental Embryology and Development. Unfortunately the intellectual and philosophical aspects of man's story were often kept as well guarded secrets or esoteric knowledge of a restricted number of initiated members of the priesthood; in the course of time most of the oral and written records perished, victims of changing times and persecutions. Thus the teachings of Hermes Trismegistus of ancient Egypt that could have given us an insight into the evolution of man's intellectual processes, were only partly revived by the Greek "mysteries", mainly at Delphos and Eleusis, but they have come down to us only as a few scattered references that do not permit a reliable representative reconstruction of the body of knowledge of that school of thought. The *Veda* are the only body of ancient knowledge that have come down to us as an extensive and cohesive whole through an unbroken chain of teaching, both oral and written. This immense body of knowledge can, even at this late stage, help us to form an idea of how the human intellect and religion evolved, at least among the Āryans and perhaps in a comparable manner among most of the Indo-European people, and perhaps others as well.

It is therefore in the *Veda* that I have tried to trace the origin of Man's beliefs. The remarkable feat achieved by the Āryans who migrated to India of memorising the whole of the *Vedic* lore, generation after generation, brings forth almost intact the thoughts of ancient man. The oral tradition and later the written transmission of the *Veda* covers such a long period of time that it allowed for enrichment of the primordial thoughts with later findings. The personalities of the Gods themselves evolve as knowledge develops and as everything else in the universe evolves, throughout the *ṚV* itself, as Dandekar has shown (1979a, p. 142). Knowledge was gradually made accessible to more initiates and it percolated in part to the common man by the addition and interpolation of mythological tracts as well as by adding empirically acquired knowledge of the phenomenal world. It therefore appears to me that the *Veda*, as we have them, mostly have:

1. a literal meaning;
2. a figurative or esoteric meaning;
3. a magico-divine-spiritual-religious purpose for propitiation of the divine and of each divinity through chanting and incantations to be used mainly or exclusively by the initiated *brahman*.

Throughout the *Veda*, specially the *ṚV* there is a constant questioning and unremitting search for answers made by men who try to understand what goes on around them (ex: *ṚV*. I; 161). Consequently my question was, would traces of the basic conceptual ideas be identifiable in the *Ṛgveda Saṁhitā*? Since my basic hypothesis is that original ideas can only be derived from the observation of the working of biological and ecological principles of general validity and since the *Vedic* and Hindu traditions are remarkable in their ecological visions, in them indeed is to be found the original *forma mentis* of Diogenes' "naked biped".

It should be pointed out that this approach is radically different from the often quoted, though now outdated, school of thought that regarded the Gods of the *Veda* as personifications of natural phenomena. What was indeed divinized by the ancient ṛṣis were the identified and rationalized relationship of cause and effect and the scientific principles that the ṛṣis gradually discerned. The II, VI and especially the VII *Maṇḍala* of the *ṚV*, the Vasistha hymns, can be read and are to a great extent, scientific books. Much however, still remained unexplained and could not be "represented mentally"; all that was then relegated to the realm of magic, as amply documented in the *AV* that unlike the *ṚV* is practical rather than speculative. Some of the Gods, and most important among them Indra, the **King** or **Chief** of the Gods, were historical figures deified, a practice common to all people who at some point or other of their cultural evolution have gone through a stage of "hero worship". Even nowadays, in at least one developed country, Japan, the divinity of the Emperor, on which the ancestral religion is based, is being seriously discussed. Dandekar analysed in detail the evolution and deification process of the personality of Indra (Dandekar, 1979a). Incidentally it could be pointed out that the principal deity of the Kafirs of Kafirishthan (now Nuristhan) who are descendants of Indo-Europeans, probably Āryan stock, and similarly the Kalash of the

Gilgit-Hunsa area, still have as their main God **Imra** (or Indra) for the Kafirs and **Balumain** for the Kalash, who arrive at festivals on horse-back, or who are represented by wooden carved horse heads, though the people themselves do not have horses any more. The horse is one of, if not the most important and representative domestic animal of the Indo-European people.

The process of direct acquisition of knowledge is described in many hymns, specially in *ṚV, Maṇḍala* IX, all of them dedicated to *Soma Pavamāna*. Many of these can be read as perceptions acquired during wake-up observations (*ṚV*. IX; 95-96), others could be read as descriptions of supra-sensory perception perhaps induced by the use of different plant-extracts; *soma* is, at places, mentioned in the plural. Supra-sensory perception could also be reached by self-induced hypnosis and deep meditation. All of these are well known techniques to acquire enhanced perception, to control body functions and to bring the unconscious and the sub-conscious to the fore at conscious level; "born out of time as finder of treasures" (of science). . ."king of all existence. . .as *Soma Pavamāna* you have been flowing to the earliest man. . ."who sought knowledge.

Once the relationships of cause and effect were conceptually grasped, the understanding acquired had to be mentally crystallized, so to speak; into on entity above suspicion: a God. But since nothing in the phenomenal world is ever fully understood, nor can any event be ever forecast exactly or even anticipated accurately in all details, except for astronomical cycles, this Entity is much above man; **It** obviously knows and understands Ṛta and all that man is unable to fully grasp mentally. God is implicitly omniscient and of course present everywhere and at all times; **it** is omnipresent, because phenomena are in the process of being and becoming, everywhere, always. Man accepts that he is not nor will ever be capable of being omnipresent and omniscient. This scientific attitude and logical frame of mind is the golden thread that runs through the *Veda*. The ancient seers kept their mind open to recognize and admit new knowledge as understanding progressed. They then incorporated the new knowledge making use of their power of reasoning and discrimination. Evolution of everything, including of the personalities of the Gods was thus implicit. In fact, ideas and concepts and science itself evolves as knowledge progresses.

I believe that the first God, primordial idea of God was Agni (fire, heat, energy) who underlines all phenomena of the living and non-living. Ṛta could not be divinised, because it is the recognition of a *status quo* and not a concept. *Ṛta* **Is** and has to be submitted to, obeyed, preserved, analysed and known as far as possible; it is better to abide by it because it is how things are and because first of all Ṛta determines one's place in the universe, one's duties and one's functions and therefore one's *dharma* as well as one's own rights. Vāk also could not become a God because Vāk is a means to an end: the transmission of an understanding. Vāk is the means by which mankind and all domesticated animals and plants benefit from the studies carried out by the ṛṣis. The contemporary counterparts of Vāk are scientific and educational journals, books, teaching aids such as photos, posters, maps, films, videos, TV and so on.

Rta and Vāk are not Gods, though they are respectfully deified, they are seldom personified or even represented iconographically, they are often called **ancient ones** and they call for the utmost obedience; they are present, though often indirectly, throughout the *Veda*.

Ṛta is the result of the cosmic sacrifice of creation (**Sacrifacĕre**, to make sacred) *ṚV.* III; 6, 6-10) at which Lord Agni is present to provide all Gods with "noble worship". Agni is unconcealable and kindled by all the Gods (ṚV. X; 64, 3 *et al.*) with whom he is permanently associated. Gods are clearly differentiated from natural elements that are listed later in the same hymn (*l.c.*; 64, 3 *et seq.*) as rivers, floods, trees, mountains to **conquer** which Agni is again invoked. But Agni is not invoked only figuratively, he is also invokėd physically and practically to vanquish enemies, animals, men, demons or wicked spirits. The destructive power of the jaws of Agni, (*ṚV.* X; 87), the flames, are repeatedly called to aid; Agni is also called to be destroyer (*l.c.*; 12) of the "fool who ruins truth with falsehood".

While studying the *Veda*, I made three assumptions:

1. The *Śruti* are stated to be revealed truths, consequently it would be impossible to find in the texts the explanations of how the knowledge stated therein was acquired. We either accept it as a revelation or we assume that the knowledge-acquisition process was kept secret purposefully. In either case it will be impossible to find out from the text **how** the knowledge was obtained; we can only put on record **how far** the understanding had progressed at the time when the *Veda*, especially the *ṚV* were formulated; the *ṛṣis* sought Truth that "is the basis that bears the Earth" (*ṚV.* X; 85, 1). There are, however, "two truths, the apparent and the one that is concealed. . . known only to those who are skilled in highest truths" (who, in my reading, are the initiated few) (*ṚV.* X; 85, 16). Furthermore, the ancient *ṛṣis* were primarily concerned with physical and spiritual or divine Truth that are the matter of fact of nature, rather than ethical or sentimental truth that are human expressions.

2. The *Śruti*, as we have them, are a **transcription** of the ancient oral tradition. This necessarily had to be concise in its expression. When the oral tradition was written down, similar to what happens with the transcription of all ancient scholarly texts, the language used had to be a sort of code, because:

(a) pen, pencil and paper were not available and writing material was scarce, cumbersome and limited; records had to be brief and easily memorized;

(b) aphorisms, metrical poetry and music are obvious means for correct and easy transmission of ideas impeccably repeated; they serve as captions and allusions to best recall the oral teachings of the *guru;*

(c) secret knowledge can be imparted to only a few carefully selected or initiated, tested and trusted *śiṣyas.* Esoteric language is often purposefully enigmatic and has to be interpreted and explained by the *guru.*

3. The *Śruti* contain the **basic scientific speculations** and conclusions obtained by observation, logical deduction, empiricism and experimentation, and they are rendered by figurative and symbolic imagery.

In conclusion I realized that most Gods represent ideas, concepts and laws, that altogether make up a body of knowledge that relates to different aspects of the inner and outer world of man.

Science, Art and the *Veda* are fundamentally similar in that they describe and communicate the meaning and function of the many aspects of our inner and outer world, but they differ in the means and ways by which they communicate their insights. Typical scientific-bio-ecological aspects are those with which I am conversant and an expression of how they are expressed is clear in *ṚV*-VII; 76, "Dawn illumines the path to be trodden by the sages and all men long for dawn", where dawn is what precedes sunrise: it is the beginning of "seeing" or understanding for which men are longing; dawn is what comes before knowledge and "light unveils the darkness of ignorance" and puts in motion life and the food web (*ṚV*. VII; 75). Dawn is (*ṚV*-VII; 80,2) youthful and unquestioned . . . she has "turned thoughts to the sun and fire" of life.

"Thus the fathers could find the light that lay in darkness" (*Ṛ*V. VII; 76,4) and "the ancient sages true to law eternal were the God's companions at the banquet". Sage Vasiṣṭha (*Ṛ*V. VII; 87, 4) also reminds man that there is still much yet to be known. More specific examples I will give in the chapter on the bio-ecological aspects.

I shall discuss Lord Agni in more detail later, Agni who, as God, is not the same as common fire that appears in multiple forms. The antiquity of Agni as concept of energy and as God is further attested by the fact that it is seldom personified and represented as such. Visual and personified representations of Gods came later and they were preceded by a long period of symbolic representation in graphic, sound and totemic form. Stylized or symbolic representations of fire are ubiquitous throughout the world to this day, in all cultures.

After Agni, the original God, perhaps the first concept grasped and deified by man was *Savitṛ*, the great Vivifier, Life and Hope who is represented as a struggle; to win the struggle one needs the determination provided by Agni to acquire wisdom and power (*Ṛ*V. I; 66). The concept of Life and motion, and perhaps also of Time that is inseparable from Life and motion is deified in the Great God Savitṛ.

The antiquity of Savitṛ is mentioned over and over again and is further confirmed by the fact that there are few representations of Savitṛ, which means that by the time iconography was developed, he was already so ancient that he was implicitly accepted as ever-present, as embodied is the Sun that is the giver of life on Earth; at that time new divine personalities would have developed and were represented iconographically. We could, however, speculate that the *svastika* and the *cakra* that is derived as a symbol from the *svastika* is a symbol that occurs throughout the world in many different cultures. Incidentally it could be mentioned that most *svastikas* are represented forward moving, from left to right, indicating progression for right handed persons and for those writings, as Sanskrit that proceed from left to right. The Nazi *svastika* was turned the other way round, to the left; was it an unwilling and unwitting recognition of regression and degeneration, a significant Freudian slip?

Man, the apprentice sorcerer who fears his own inventions (*ṚV.* I; 128, 2), can control fire made by himself, except as mentioned repeatedly, if it runs away as forest fire or grassland firs, but Indra's thunderbolt cannot be controlled by man (*ṚV.* I; 120, 10).

In conclusion, the *Veda*, read with a biological-ecological decoding key show a profound knowledge of the relations of cause and effect of natural phenomena, specially those that are related to life processes. The *Veda* are a treasure-mine of information about the science of ancient man and, even more important, of the way man thought, understood and transmitted his knowledge to others.

"May the most glorious *Pitṛs* aid us. . ."

ṚV. I; 106, 3

3

The Concept of God

"Agni, thou encompassest the Gods as the felly encompasses the spokes of a wheel"
ṚV. V; 13,6

TWO elements appear to be indispensable to visualize and then formulate mentally the concept of God; first, the existence of a powerful mind incessantly intent on understanding natural phenomena and second, the mental capacity to formulate or give a body or formula to abstract ideas and concepts. This non-material body is what was later called the **forma** or form. The second step is an immense leap forward in the evolution and development of the human mind.

All of us are imbued with the concept of God. To us God is no longer just an idea but is a mental reality whether modern human beings believe in **It** or take great trouble to dispose of **It**. Implicitly it appears natural to us that the idea of God existed throughout the ages of man's mind. This, however, cannot be so and at some stage, either by revelation, by inspiration, by a "stroke of genius", by a sudden thought or idea, or whatever we want to call it, or by step by step thinking and logical deduction, as for instance expressed in the *Īśā-Upaniṣad,* man must have conceived of the idea of an Entity, invisible yet present, intangible but that can be felt, omniscient, omnipresent and omnipotent, with no beginning or end in space and time, better known from the effects of its existence than known in itself. Indeed it is stated that the Supreme Being is so formidable that man cannot appprehend the concept in its entirety. What man could understand of it was expressed as knowledge of the world he lives in; the whole world that is sacred because it **is part** of the Ultimate Reality. And what man could not understand, that too was inherent in the concept of Ultimate Reality or God; thus man, to appease his eager mind, had to create **approximate** understandings and representations; they appear in legends, myths, magic and visual representations. Sound itself is made a revelation or *Śruti*, as the syllable **AUM**, ॐ that sets in motion all vibrations. Nature and magic are inseparable and magic is as old as mankind, and of course it includes prayers, incantations, exorcisms and shamanistic practices that are all part of the acts or rites that man developed to satisfy his urge of active participation in the Cosmic Drama and Cosmic Order. These acts are included in the power of *asu* Varuṇa, so well analysed by Dandekar (1979a, pp. 28-67; 312-331) and are incorporated in the rites and ritualistic performances of ceremonies, religious and laic (*ṚV.* I; 110). The concept of God and Agni as God is expressed in *ṚV.* VIII; 11,6: "Sages, we call the sage to help; mortals, we call God to aid, we call on Agni with our songs" (transl. Griffith).

Once it was recognized that there is something unexplained and therefore mysterious in this world and in the universe at large, something that is beyond the grasp and control of man, faith, religion and dependency on this entity were born. Man had to develop rituals to bring himself in closer relation to his God; rites and rituals were probably first created to propitiate the invisible Omnipresent. At the same time it must have been recognized that the role of empirical and experimental science, though not specifically described as such, was of fundamental importance for the survival of man who is not particularly gifted physically. The *ṛṣis* were good scientists in their own right; not scientists as we visualize them today, going around in white laboratory coats and dealing with sophisticated equipment, but scientists in the full acceptance of the term, meaning "those who understand and know; those who endeavour and work to acquire understanding and knowledge" of the inner self, of the outer self, of the visible and the invisible, of the tangible and the intangible, of the past and future. The VII *Maṇḍala* of the *ṚV* can most of it be read as a book of science culminating in intense devotion to *Asura* Varuṇa. In other terms, Varuṇa of infinite magic power, takes over when and where man can no longer understand and explain nature with his own limited intellect. Dandekar (1981, pp. 113-121) discusses vividly the *bhakti* aspects of the devotion of Vasiṣṭha to his God, Varuṇa.

Prof. Dandekar (pers. com.) objected to my reading **science** in the *ṚV* saying that no scientific theories were developed by the *ṛṣis*. This is correct in the contemporary acceptance of the expression **scientific theory.** Theory is defined here as a body of doctrines used for the explanation and prediction of a phenomenon or a series of phenomena. Though the ancient *ṛṣis* probably did not see the need to knit together the knowledge acquired into a well defined and formulated whole or **theory** having general validity, they did recognize the importance of factual knowledge and of the establishment of cause and effect relationships for the understanding of the phenomenological world. They based their understanding on observation, logical deduction and experimentation. Maybe they did not formulate theories because they did not see the need for generalization since the particular explanation was sufficient to the purpose; or perhaps they did not have sufficient information to generalize their understanding. Most probably the concept of **theory** did not arise in the mind of the *ṛṣis;* the intense devotion that Dandekar rightly recognizes in *Ṛṣi* Vasiṣṭha is indicative that *Asura* Varuṇa is the embodiment of all understanding and that he himself and He alone is the explanation of the

Universe, since He is the *asu* of the universe, the infinite magical power that permeates everything originally Varuṇa is himself **the** theory.

The establishment of a theory requires taking a big step from the particular to the general. While this is relatively a direct procedure in the exact or mathematical sciences, it is a very complex one in physical and chemical sciences and is extraordinarily more so in the biological sciences where the number of factors and variables is practically infinite. The world of the *Vedic* man is essentially a biological and ecological world, therefore one whose mysteries are difficult to unveil. Incidentally the word theory can be derived from the same root as God in Greek: *Theo*, and would therefore mean: "what pertains to God". Consequently all knowledge is the science pertaining to the *Deva*, same root as *Deus* in Latin and *Theo* in Greek. I believe that the *ṛṣis* progressed in a genuine scientific manner as far as they could go, but where they could not give a clear answer they had to explain the world as "God's magic".

The great achievements of the ancients as they have come down to us through teachings of the *Veda*, are: 1. a critical attitude that stresses the power of discrimination of man as different from animals; the power of discrimination enables man to identify what he knows and what he does not know; 2. the recognition that everything in the cosmos is subject to perpetual evolution, that there is a conti nuous change over time of the whole and of its parts, so much so that even the personalities of the Gods evolve; 3. the recognition of the cyclic aspect of natural phenomena as perceived by man living in an Euclydean world and that these cycles can be broken down by cataclysmic events.

Coming back to the very early man that had just emerged as *Homo sapiens* and who had learned to **use**, then also to **make** and **control** fire, we can easily imagine him to be quite bewildered as the apprentice sorcerer he really had become. Fire indeed has all the attributes to become the precursor God and by a process of identification and assimilation, God himself; witness to this is the sublimation or disappearance of matter caused by fire. Fire, **the** "*ur-gott*" of thinking man is lovable and fearful, gentle and terrible, destructive and generous, it takes and gives, it is good and bad, above all it is powerful and omnipresent, great, imposing, giver of light and warmth, giver of life and food and never fully understood since it is ever somewhat mysterious. We have all experienced at some time or other the hypnotising power of fire. Man learned to kindle fire, but he could not similarly kindle the attributes of Agni, the "fireness" in fire: "Agni bring hither Agni's self" (*ṚV.* I; 44, 7-8), such as for instance the fire of life. Primitive thinking man must have had the idea and must then have developed the concept that the entity, God, that he could not really define or understand, **expresses** itself through fire, more than just **manifests** itself. It expresses itself by what it can do and by the uses it can be put to. That which fire consumes disappears into nothing and this power of sublimation makes Lord Agni the Purifier par excellence, and, later on, the ideal intangible, spiritual messenger between man and all Gods that were born later in man's mind and that exists everywhere, in space, air, earth, water and within man himself as thought and as the fire of life. "He alone of all these Gods . . . hadst been born living from the dry wood; then all were pleased with thy wisdom" (*ṚV.* I; 68,2) . . . "within the laws of Ṛtā" (*ṚV.* I; 70, 2). Agni grants spiritual vigour and gives immortality to the Gods (*ṚV.* I; 3;4), "the Gods gained immortality from Agni, at first only Agni was immortal".

I concluded that partly because not everything about fire could be understood by man, fire was developed into a God, Lord Agni who is the essence of fire, who appears as flames, heat, warmth, energy, sun and who hides away in waters and wood in his **potential energy** form, but not revealed aspects. In all the manifestations of Agni there is a common principle; he is omnipresent as energy (*ṚV.* I; 27,2), unexplained, undisclosed, mysterious and adorable.

These aspects of fire are aspects of God, because by definition God is never totally comprehended or mentally visualized. The kindling of the innermost fire for one's enlightenment is well described in *ṚV.* V; 2. The original and supreme God cannot be fully represented mentally because of human intellectual limitations; unlike God, man cannot unravel all mysteries. This is nowhere expressed better than in the simple common prayer:

"O Lord, forgive three sins that are due to my human limitations.
Thou art everywhere, but I worship Thee here;
Thou art without form, but I worship Thee in these forms;
Thou needest no praise, yet I offer Thee these prayers and salutations;
Lord forgive three sins that are due to my human limitations".

As God Agni "by whom the sages gained their wisdom" (*ṚV.* IX; 108,4) and through whom the wish is expressed: "may we accomplish all that we are capable of" (*ṚV.* X; 2,3), came to be mentally realised, the need arose to invoke, propitiate, beg and adore the mysterious entity that was forced on man's mind by his own reason (*ṚV.* I; 19) and was expressed through Agni; rites and religions were born, hence also the appellation "divine fire" that becomes more and more frequent. Fire divine as knowledge and wisdom; knowledge as Truth. Truth is the obsession of the Āryans (*ṚV.* I; 71 - *ṚV.* I; *72* - *ṚV.* VII; 66, 13) as is also evident throughout the history, legends and myths of most Indo-European people, as for instance the Celts, to mention only one group. The inner fire of consciousness later became the fire of *tapas*, as in *ṚV.* I; 73,4. Fire is also the source of life, physical and spiritual but according to the eternal duality of everything when one takes a coin, one accepts the other face as well. Fire creates life and also destroys it.

The sublimation of Agni into God is a good example of how abstract ideas rather than physical phenomena were made Gods and, in my opinion, it is the best example that the *Vedic* Gods are not just the deification and personalization of natural phenomena, but are instead the representation of **philosophical principles** and concepts. In other words I believe that ancient man crystallized his knowledge into Gods, though he also deified successful leaders, chiefs and teachers. Figurative myths and legends are useful instruments to instruct, other men and perpetuate historical facts. Gradually the scientific, historical, mythological and existential experience of the early Āryans who migrated towards Saptasindhu were condensed into that formidable body of knowledge that are the Śruti.

In historical times, the details of the evolution of the Buddhist sects, for instance the *Jātaka* stories, show clearly how mythology develops **after** and not before the philosophical religious aspects of the doctrine have been established and how they adapt to the conditions of the pre-existent cultural environments.

Radin (1941, pp. 203-203) says that "the Australian Aborigenes, the Andaman tribals, Negritos of the Philippines, the Lalkman of Tierra del Fuego have no myths of creation, the Supreme Being is not considered a Creator . . . no prayers are addressed to him, and there are no rituals associated to his cult"; further Radin adds: "there is no religious belief properly, though only **he** is always good and just"; "there is *only* a philosophical notion". By saying this Radin indirectly expresses the same views as given here that "philosophy" comes to the mind of man **before** religion.

Humility is the characteristic of the early stages of religions (*ṚV.* VII; 89); arrogance in man likening himself to God and considering that it is God's duty to satisfy his wishes, desires and prayers, as well as representing God in man's image, come at a later stage.

One is tempted to read the origin of the concept of God directly at many places of the *Veda*, for instance in *ṚV.* I;

65, Lord Agni is said to be born in the womb of Ṛta; in other words, the idea of God is born from observing nature's and all laws of the universe by which man must abide (*RV.* IX; 63, 14-24). In *AV.* X, 2, man the observer examines himself and the outside world, he finds out how wonderful is the anatomy, the structure and function of the organs and systems of the human body and he exclaims in awe: "the Gods did it", "the voice, the brain, the thoughts, the pleasure", "the Gods did it". All structures and functions are analysed; then the final great question comes naturally; it expresses the idea of God and formulates the conclusion of the existence of God: "By whom, was the mind put in him? Wherewith does he dwell upon the Gods?" and the concept of fire as energy is at the root of all thought. Agni as thought, body heat and energy is the expression of the invisible and intangible God.

Traditionally the significance of fire worship is to imbue wisdom in man: "Agni impart to me wisdom, *brahman* impart to me *brahman*" (*AV.* XIX; 11,4). Agni as the fireness in fire is knowledge, wisdom and mental power, but fire also burns dirt and confusion (*RV.* I; 79, 12) physically, spiritually, intellectually through the power of discrimination that is his gift to man. Fire burns evil thoughts and provides energy for keeping and acquiring more knowledge. There is a clear distinction (*RV.* I; 79) between Agni the spiritual and intellectual entity and "Agni of the pointed flames" (*l.c.*, 10). However, both Agni and common fire are, like everything else in ecology, closely related to other Gods as is evident from the frequent recurrence of the duals Agni-Indra; Agni-Mitra; Agni-Soma; Agni-Varuṇa and Agni the lofty *Asura* (*RV.* V; 12), all pervading and present everywhere at all times (*RV.* I; 149; 151). The individual as well as the collective society, need to have their internal world attuned to the external world of which they are a part to achieve ecological inner (psychological) and outer (physical well-being) balance. Accordingly there is a need for "seasonal offer of the inner sacrifice" (*RV.* I; 170, 5), which is a renewal of the self in tune with the seasonal changes of the environment over which man has no power. These are phenomena that can only be influenced through magic or religious practices performed preferably at the solstices or equinoxes. Usually the winter solstice marks the beginning of the year and the spring equinox is specially significant as marking the re-awakening of nature after the winter stasis. In this context, in addressing the sun, the horizon is called "the lap of your mother Earth", which is more than just a poetical expression. Natural cycles are a recurrent theme of great importance (ex: *ṚV.* V; 29, *ṚV.* I; 105, 9).

In the *Veda* the concept of Dhātṛ as Creator must have evolved naturally from the fundamental triad Agni, Ṛta and Vāk. If Dhātṛ is indeed the idea of a Creator or Maker, rather than a fully developed divinity, I would place him at a very early stage of the development of the Āryan man's mind when the gradual acquisition of knowledge imposed the logical assumption that "what **is** had to be **made**".Dhātṛ as an idea, rather than a God, belongs to philosophy rather than religion, in fact, in the evolution of the *Vedic* thought, from philosophical to religious, Dhātṛ gradually fades off and becomes a secondary divinity. On this point I would not follow the conclusion (Dandekar 1979a, p.16) that "abstraction . . . divinities belong to a later phase of the *Vedic* religious thought". I would repeat here that I believe that pure abstract concepts are at the root and precede religions; elaborate thoughts evolve into religions, rather than the other way round.

I mentioned above the Āryan obsession with Truth that is strong to this day in all forms of Hinduism. It could have originated from the fact that originally the two concepts of Truth and Knowledge were interchangeable and were used as synonyms. Knowledge is power, hence the need to acquire it and in so doing, Truth becomes sacrosanct.

Once early man developed the concept of God, the knowledge of God would have been the ultimate goal, because man recognized the "magical potence of knowledge". Dandekar (1979b, pp. 149-161) stresses the point of how this fundamental trait was retained throughout the history of Hinduism as "respect for knowledge, and trust in the practical religio-philosophical efficiency of knowledge". "Do not forget, O Agni, who art a Sage possessed of knowledge" (*Ṛ̣V.* I; 71, 10).

Finally I would also mention that it is made clear at a few places (ex: *Ṛ̣V.* I; 164, 10) that some aspects of knowledge must be kept secret, therefore knowledge of the "mysteries" should be given only to a few selected initiates. Complex rituals, sometimes performed secretly and elaborate permanent structures such as temples, churches, places of worship are constructed. They are useful to explain the doctrine to the masses that must also be impressed and filled with awe, respect and admiration. Rituals and impressive permanent structures may also be an unconscious attempt to capture and retain God as an honoured guest.

"When did I worship the deities of falsehood
  or when did I think adversely of the Gods, O Agni?
Why are thou angry with us, O *Jātavedas?*
  May destruction fall upon those who lie against thee"

Ṛ̣V. VII; 104, 14 (from Wilson and SS & SV transl.)

4

Ecological Aspects

"He (Agni) who gives food, like patrimonial riches
  and guides aright like some wise man's instruction,
Loved like a guest who is received with respect,
  may he as *Hotr*, bring prosperity to the house
    of his worshippers"

Ṛ̣V. I; 73, 1

"Weal to thee be Agni, together with the waters,
  Weal be *Soma*, together with the herbs"

AV. II; 10, 2.

THE ecological validity of the *Veda* is not exclusive to them, St. Francis of Assisi expresses much ecological wisdom in his "*Cantico delle Creature*" (Canticle of all Creatures); Chief Seattle in his letter to Franklin Pierce, then president of the United States of America, is detailed and explicit in his defence of the environment and he bases his arguments on valid ecological principles; he declares beautifully how he and his people are an integral part of their environment: "how is it possible to buy or sell the sky or the heat of the land?. . . as we do not own the air or the lustre of the waters how could they be bought?. . . each small piece of soil is sacred to my people!".

Books have been written and more could be added about the mystics, Rumi and the Sufi tradition, the Christian, Jain, Buddhist, Tibetan as well as some of the Chinese and Japanese poets and even the great mathematician Omar Khayyam and other Iranians whose views on the permanence or impermanence and cycling of all matter and energy are ecologically sound in modern scientific terms. Giordano Bruno could be revived from the stake for his "heretical" views, but I will strictly adhere to my first purpose. The Veda, as we have them, constitute a considerable body of knowledge, codified in rigid metrical form and collected in the *Saṁhitā*. In fact, I started with the aim of studying the ecological aspects of the *Śāstra* at large; by gradually trimming down my ambition, I am at present concerned only with the *Veda*. The wealth of information I see in these would be enough to fill a lifetime with assiduous study, while my own life time is quickly approaching its end.

The *Saṁhitā* are the product of a highly evolved and evolving culture; a culture that to this day has kept alive and creative the memory of its roots and beginnings. The critical ideas, thoughts, concepts, conclusions and technicalities of the ancients are present in every verse, crystallized as myths, pronounced as teachings, rendered in poetic form, memorized as sound and metre, spiritualized as invocations and laudations, dictated as techniques. All these aspects are grafted on the ancient roots of a system that preserves as sacred all knowledge acquired and keeps its doors open to any useful addition "from the four quarters".

Indian tradition admits three distinct ways in which the *Vedic* hymns may be interpreted. The inner subjective way (*adhyātmika*), the objective literal or material way (*adhibhautika*) and the godly or spiritual way (*adhidaivika*). What precedes is mostly my personal subjective way of reading the hymns as a biologist. What follows is an objective interpretation from an ecological point of view. Saraswati and Vidyalankar (1977, SS and SV) have done a beautiful work in rendering the *ṚV* in an integrated manner, with emphasis on the *adhidaivika* aspect as well as the *adhibhautika*, their mounmental work has been the beginning and basis of my studies. Of course, I would not add anything to the learned expositions of the commentators, specially on the *adhidaivika* aspects. For instance, under *ṚV.* III; 35, 8 they write ". . .extraction of medicinal elixir from herbs. . .needs an expert . . .pressing stones. . .and water, and is sweetened with milk. . .on the metaphysical plane, the worshipful sentiments are sweetened with intellect, austerity (stones) and pure thoughts".

Traditional way of life

When reading the *Veda* from a biological and ecological point of view, one should alienate oneself from the contemporary world and environment. The best way to visualize how the *Vedic Ṛṣis* saw themselves, their fellow men and nature, is to figure out how man lived three of four thousand or more years ago in the areas inhabited by the ancient Āryan tribes and before them by the local populations whom they collectively called Anārya and with whom they came in amicable or not-so-amicable contact. It is indispensable to try to identify oneself mentally with the world as it was at that time to avoid such elementary mistakes as for instance taking tree-worship as worship of the sacrificial post (*yūpa*). One must imagine how impressive must have been the giant trees of the extensive forests of those distant times and read *Vanaspate* literally: "O Lord of the forest"; "O tree". When in the text of the *Veda* sacrificial post is meant, it is thus stated. The "interpreted" erroneous translation of tree or lord of the forest into *Yūpa* is just one of the many examples that indicates how distant the learned scholars who worked in their libraries in far away places were from the environment and realities of everyday life of the *Vedic* people: "*Vanaspate*, O Lord of the forest, give fourth thyself and call the Gods for sacrifice". . ."May Agni, God intelligent, speed our oblations to the Gods" (*ṚV.* I; 142. 11) where the tree is invoked as the wood that will sustain the fire through which Lord Agni expresses himself. Tree worship is a very ancient practice still alive in one form or another in most cultures; it is particularly significant in present day Hinduism and Buddhism which have carried it over respectively from the ancient forms of Vedism and Brahmanism. It is recurrent in Palm Sunday's waving of palm fronds and olive branches, as the "tree of life" of Christian texts, as the Christmas tree and other trees in the Christian tradition that still maintains some products from trees as important elements in the liturgy and mythology, such as incense and the symbolism of the olive branch of peace brought by a pigeon. Incidentally the pigeon has no gall bladder, this could be one reason why it was taken to be a non-bilious animal and was chosen as a symbol of peace.

The South-Siberian Turks consider the *axis mundi* to be a tree and the shaman's drum is made from that tree (Marazzi, 1984, p. 11); they also pay homage to the "rich tree of the Altai top" (invocation to the *Altay Azi Akalar* Spirit, *apud* Marazzi, *l.c.*, p. 110). Tree worship is still a daily practice for devout Hindus.

The migrating Āryan tribes were mainly given to pastoral activities and only secondarily and in a small way to agriculture. Migrations took many centuries as they stopped at convenient places for shorter or longer periods of some years or some generations. The best pasture or agricultural lands would be disputed away from earlier colonizers if any were present and the prize of land was the cause of disputes and wars between the Āryan families and clans themselves, but eventually a strong desire to settle down grew among the people on the move (*ṚV.* VI; 49, 15): "give us a lasting home" and "one rich with nature's God-given bounties".

It may have taken something like three thousand years before the tribes coming from Central Asia east of the Caspian Sea settled down in their respective territories, but as far as one can estimate from the present day life in some remote areas, life styles would not have changed in any fundamental manner since then. Loude and Lièvre (1984; 1987) lived for two years among the Kalash of Chitral and Gilgit and before them Robertson (1887) lived for one year and visited several times the Kafirs of Kafiristhan or Bashgal. Their observations and those of others are invaluable aids to help us visualize how the *Vedic* man lived. Several anthropologists studied the habits and customs of Dards, Kafirs and Kalaṣh people, who are all descendents of ancient Indo-European, perhaps Āryan migrant tribes.

Vedic people were one with nature (*ṚV.* VIII; 58, 2-3; *apud* SS & SV, a *Vālakhilya* hymn): "One is that which manifests in all", which in contemporary ecological terms is expressed as "everything is related to everything else". It was essential for the survival of ancient people to learn how they could interfere with the world around them to improve quantity and quality of food, shelter, health, hunting and fighting potential. Man had to recognize what powers of nature he would be unable to control and he was thus compelled to resort to magic and religion to win the cooperation of winds, rains, regularity of the monsoon and astronomical recurrences, for the control of earthquakes or natural dam-bursts, forest or savannah fires, river flow and all major elements of nature. All these are described, invoked and prayed to in the *Veda*, with fear and respect. Interference of the Gods was sought for the repetition of favourable phenomena such as regular rains to ensure rich pastures and good crops, healthy cattle, horses and sheep, as well as fecundity of the land and of women. Rites would be performed also to forestall destructive natural events. Man learned that there were also natural phenomena with which he could interfere and with which he could collaborate after having understood how the **system** worked. Thus rational agriculture and the use of natural products such as provided by plants, or the smelting of metal ores and improvement of the wheel, the plough and irrigation devices as well as harnessing of horses and bullocks were already well developed in

Ṛgvedic times.

The seeker of nature's laws is rewarded: "the fathers found the light (knowledge) that lay in darkness (ignorance) and with effectual words begat the morning (teaching of knowledge, or acquired science) (*Ṛ*V. VII; 76, 4). In fact, we would not be here if our forefathers would not have been wise and ingenious.

The intimacy of man with nature is expressed with great joy (*Ṛ*V. VIII; 31, 10) "we solicit the happiness afforded by the hills, the rivers, the sun. . .". "Agni the first God. . . who makes our fields prosperous" (*l.c.*, 14). It is clear from *Ṛ*V. X; 85, 1-19 and the interpretation given by SS and SV mainly based on the *Nirukta*, that many rites and rituals derive directly from the observation of natural phenomena and cycles, and not the other way round, the phenomena are not **explained** by the rituals.

Communication of knowledge

The *Ṛṣis* who retreated to their *āśrams* to study and apply the fruits of their studies to practicalities had ample time also for meditation and for religious practices and thus became the famous *ācāryas* who imparted theoretical, spiritual and practical knowledge to such youngsters as were eager to learn. Apprenticeship lasted for many years of unremitting, arduous work under trying conditions aimed at strengthening the body, the mind and the will, thereby elevating the spirit. Evidently, the concise texts of the *Veda* reflect only the essence, or the conclusions of the vast amount of empirical and experimental scientific knowledge that accumulated generation after generation of *ācāryas*. Physical, intellectual and spiritual abilities were transmitted practically and were impeccably memorized thanks to strict discipline, to yoga practices and to metrical (rhythm and sound) laws without written transcription; memory was also aided by an extremely rich mythological lore that is still growing through addition of new myths or the addition of details to old myths. There are, for instance, several traditions on how the right (or left) tusk of Gaṇeśa was broken.

Material Culture

In their migrations from the steppes of Central Asia through Balkh in Khorassan (Dandekar, 1979a, 1981) the ancient Āryans came in contact with many new environments and different materials. The wealth of the rich valleys of Saptasindhu inevitably favoured the development of many new abilities "with the mind of a seer and the acts of a sage" ("*Ṛṣi mana ya ṛṣi krit; ṚV.* IX; 96, 18). During earlier times the most important objects of daily use would have been those of pastoral people, in fact in a hymn addressed to Pūṣan (*ṚV.* VI; 53, 9), the guide, the path finder (Dandekar, 1979a), the "goad with a horny point" is mentioned. Though it is not certain that the word *gūpaśā* means horny material, there is no doubt that leather straps, objects and containers made of hooves, horns and hides were of common use; such were trumpets from horns, chords from sinews for bow strings and reins, thongs and latchets for footwear, straps and cover for the drum, whole hides to make drinking-water bags, covers for shelters and inflated hides as floats to ford rivers. Fat and bones of domestic and hunted animals were used by all shepherds and hunters of all cultures as raw materials, as well as cow or camel dung dried and mixed with straw as fuel. The ancient *Ṛgvedic* man was no exception. In my childhood farmers in Tuscany used cowdung cakes as fuel and we liked the milk boiled on those fires better than any other "because it smelled of our friend the cow". Cattle or camel dung is normally used for making fire throughout Asia. Over the centuries, techniques were improved, for instance by curing (tanning) leather with the bark of certain trees rich in tannins, a very ancient practice that wandering people would have learned from one another. Raw leather however, is still in use in remote arid areas where fuel is scarce and there are no tannin producing trees, such as the Tibetan plateau or among the Dard nomads of Ladakh. The importance given to the tree and consequently to the forest is obvious (*Ṛ̣V.* VI; 47, 26 *et al.*) Since a large variety of useful products are obtained from them. Species diverstiy, or presence of a large number of different plants and animals must have been a most astonishing and wonderful experience for people coming from steppes and savannahs to forested areas, specially the land of the Seven Rivers. Conservation practices and judicious management are still in use; among the Kalash, live trees are never felled because they belong to the fairies (Lièvre and Loude, 1990), branches are chopped out of live trees and only dead ones are fully utilized, a practice observed throughout the Karakorum-Himalayan ranges and most of India to date. No only organic materials were used, metal ores were smelted for the extraction of metals; it is possible, but not certain that *aya* means iron, though it could mean metals in general. Smelting of iron ore and the making of steel are very ancient practices in India (Vannucci 1989, p. 131). Fire may have been used for fragmenting rocks and boulders by heating: "By the Ṛta the Aṅgirasa have broken the rock and cleft it asunder" (*Ṛ̣V.* IV; 3, 11).

Materials are otherwise taken for granted in the *Ṛ̣V* and the *SV*, and are seldom mentioned specifically, not so in the *AV* which is more pragmatic. As such, the *AV* illustrates the everyday life of the *Vedic* people, their fears, their difficulties, their aspirations as well as their material culture and the source of raw materials, better than the *Ṛ̣V.* or the *SV.* The *AV* also illustrates which were the materials used in the rites to invoke help from the divinity and which could have been used to enhance the powers of the mind to seek help by prayers and invocations. The most immediate worries of the *Vedic* man appear to be, as is natural, about food—its availability and quality, health, death, the number and health of domestic animals, progeny and marriage, but throughout there was a constant communion with nature.

It is not clear when the conch, the shell of the marine mollusc whose zoological name is *Xanchus pyraeum* came into regular use in war and rituals, it was probably late. At the beginning it must have been rare and would have been used as jewellery only by kings, princes and perhaps their *purohitas*. It is indicative of trade with coastal dwellers since early days as it was probably already in use as the conch at the time of the war of the ten kings and of king Sudās. In *ṚV.* VI; 47, 29 *et seq.* the *dundabhī* is mentioned and was correctly interpreted as the war drum, but the conch is not mentioned; it was probably unknown in very early *Ṛgvedic* times. According to *AV.* V; 20, 21 the drum is made of wood from the forest tree and headed with cowhide and belongs to all *gotras*! The making and use of the shaman's drum, used in the rituals of the South Siberian Turks who are of Indo-European extraction, are described in detail in many contemporary shamanistic chants; the practice begins with invocation to the tree chosen by the spirits and from which the hide headed drum is fashioned and procedes all the way up to the animation of the drum itself (Marazzi 1984, pp. 174-224 *et al.*).

Domesticated species of plants and animals play a key role in the life of people conscious of being one with nature. Among the animals horses, cattle, camels, goats and sheep and among the plants barley, wheat, grapes and some medicinal plants and species of grasses were important; apart of course, from the *soma* plant that is still an enigma as to its botanical identity. Trees and forests would have been worshipped as they are by all people, specially those coming from steppes, savannahs and arid places devoid of trees or having only a few scattered ones. The use of natural products would have followed along with the technicalities developed to make better use of these materials. The settlement of migrants would also favour the habit of tree planting for shadow, for fruits, for beauty and for worship.

There is an extensive literature on the ethnological, scientific and archaeological aspects of the long process of domestication of plants and animals; it is not my purpose to review the large amount of information obtained in this domain by modern science. Rather I try to visualize how the ancient *ṛṣi* saw the process of living—the phenomenon of life and how life-giving elements of nature were used.

Non Material Culture

Apart from satisfying the basic physical needs of food, shelter and reproduction, thinking man must have developed in the early stages, both the desire for immortality and the desire to learn the art and acquire the power of divination, since either would enable him to have a greater participation in the cosmic drama that unfurled under his very eyes. Both of these desires are still unfulfilled but they are not dead and they still stimulate most individuals in one direction or other. Throughout the *ṚV*, the *SV*, the *YV* and the *AV* we may read these desires openly or hiddenly declared. It was finally granted that immortality can be achieved through good deeds that will be rewarded in future lives according to the theory of *karma*, the germs of which Dandekar (1979a, p. 341) reads as early as *ṚV.* VII, the Vasiṣtha hymns. Immortality is likewise achieved through the oral tradition and by the deification of excellent personalities like Indra.

The desire to acquire the power of divination is present in people of all cultures and is present in latent form in all invocations to Agni who, as messenger of man to the Gods is also requested to bring the Gods and their spouses to man to fulfill man's prayers and thereby tell him what the future keeps in store for him (*ṚV.* VIII; 38; 39): "Lord Agni (*l.c.* 5) knows all that comes from Gods, he knows the mystery of men", while also the *Viśvedevās* become omniscient and seers of past, present and future. The desire of immortality is alive and "He (Agni) gives us kinship with the Gods" (*ṚV.* IV; 10, 8 *et al.*). Immortality can be reached by acquiring fame and is later rationalized as the fruits to which man legitimally aspires: *Dharma, Artha, Kāma* and *Mokṣa.* Intimacy with the Gods whose life ends at the end of each *Kalpa,* also enables man, who equally has a finite life span, to ask for protection and gifts.

Vedic man was technically bound to nature. There is an interesting parallel of the Greek "Gordian Knot" in (*ṚV.* VIII; 40, 6, where IndraAgni are invoked: ". . .may you cut the evils. . . like tangles of the creeping plant", the string to be cut is made of natural vines! The *Vedic* man shows his intimacy with nature as he sees health as the law and disease as the breakdown of order or law, which in this case is the break down of the equilibrium among internal factors of the body, of the mind, of their physiology as well as the breakdown of the balance between body and mind on one side and the environment on the other.

Man perceived that life is constant changes at different levels and knew that life is maintained as long as there is movement; everything evolves and develops and stillness spells death. Similarly in human society, ". . .the one to whom you give so that he may give again (to others)" (*ṚV.* VIII; 52, 6 - a *Vālakhilya* hymn); to this day the custom is observed of distributing sweets and of feeding guests and visitors at all occasions, a custom that also seals deeds and ensures observance of agreements; it is recognized as valid by law. Movement as recycling of matter and energy are observed in everyday life; everything is recycled in India, down to the minutest objects, even new materials developed by XX century technology are recycled.

A guest or visitor is God himself and is to be honoured and fed, his requests are to be granted and from him blessings are expected in return; a sort of recycling of intellectual and spiritual power.

An incidental remark may be made here; *ṚV.* VIII; 51, 1 is addressed to Indra who drinks *soma* with others, including Nipātithau; the meaning of the name is: one who carries oblations to Gods and protects the *yajamāna* who is named Nipa. A mangrove species of stemless palm-tree bears the scientific name of *Nypa* which was taken from the popular name of the plant given locally in some islands of Indonesia. All the parts of the *Nypa* palm are useful and utilized, but most important is the sap tapped from the inflorescence that produces abundant liquid that is fermented into an alcoholic drink and that can be further processed into sugar, alcohol or vinegar. The *Nypa* pam occurs in India and was earlier present along the whole coastline.

Participation of Man in Nature

The *Veda* probably took shape in Balkh (Dandekar, 1981, p. 25) at the time when the Āryans lived there before emigrating further towards *Saptasindhu*; the *Saṁhitā* form that we have now took shape at a later stage. The subject matter of the *Veda* originated from a continuous process of intellectual creation and revelation that spans a long time, probably many centuries of oral tradition. Innumerable studies and scholarly texts have been published on this subject, but whatever the historical details or texts may be, the *Veda* as we have them, specially the *ṚV*, the *SV* and the *AV* are the expression of very ancient seers who, to my mind, were also good scientists: observers of nature, technicians, engineers and medical doctors. We have the names and know much about many of these seers and the lineage of most of them is traced and known down to the present. The *ṛṣis* of the *Veda* were historical men and at least one woman, Lopamudra, even if modern critical historical analysis considers "proofs" to be insufficient. The seers/poets/ scientists who formulated the hymns lived with nature that they studied and used. A strong feeling of man's participation in the universal order pervades the *Veda;*

man is inseparable from his cosmic environment, he is not distinct from it but imbued in it. Man becomes immortal through his descendants and through his spirit that merges with the universal spirit on the death of the body. Fire as energy is present in **movable** and **immovable** natural things; Agni, the original power or God is present everywhere at all times, even in stones (*AV.* III; 21, 1) and Agni expresses and manifests itself in many forms. Energy and matter as well as the essence of God and man are one, the same and interchangeable and when bodily life ends, man like other movable and immovable things, merges with infinity or, in modern ecological terms, matter is recycled and energy is one under many interchangeable forms. In fact, even human mind is part of the cosmic soul, or energy.

At whatever place or time the songs of the *ṛṣis* were formulated, they invariably express a deep sense of communion of man with God. Nature is to be understood as friend, revered as mother, obeyed as father and nurtured as a beloved child. Nature is sacred because man depends entirely on it and because of this everything is sanctified, including man and the terrifying aspects of nature itself, such as glaciers, landslides, earthquakes, storms (*AV.* XVII) that have to be taken as they are, even if difficult to bear. Even bad people such as thieves are part of nature and have to be accepted with tolerance and their bad habits redressed with wisdom and according to *dharma*. Clearly the ancients saw that there is a relation of cause and effect in everything, even if man and the seers themselves could not always perceive all the hidden meanings, purposes and relations among the multiplicity of natural phenomena that they observed. Natural phenomena are manifestations or expressions of the Gods, **not** the Gods themselves, they are manifestations of and express the principles that govern the world and cosmic order, Ṛta.

The *Vedic* Āryans saw Truth in Ṛta "True to Ṛta, born

in *Ṛta*, strengtheners of *Ṛta*, terrible, haters of falsehood (Mitra, Varuṇa, *Aryaman*). In their felicity for best defence may we and the seekers dwell" (*ṚV.* VII; 66, 13). To this day to tell a lie is an even greater sin than to commit a theft, but *Vedic* Truth is to be understood also as Knowledge, "which is a gift of God to all men", hence it cannot be betrayed by human lies. **Knowledge** in the *Vedic* sense has much in common with what we now call **science**, hence **Truth** in the *Vedic* meaning of the word is science as was practiced and expounded at the time when the basic instruments of study were the sense organs, which includes the mind. All intellectual thinking was considered to be spiritual.

Fire and heat

Fire and heat were perceived as the central force of life and of most natural phenomena, while in scholastic philosophy the *primum mobile* is God as Love, expressed by Dante Alighieri in the "*Divina Commedia*" as "Love that moves the Sun and the other stars" (*Amor che muove il sole e l'altre stelle*). Fire as light is necessary for the growth of plants and as warmth in birds and mammals, it is linked to life and movement; cold blooded animals are inert in cold weather and a lizard basks in the sun to warm up and move. At high altitudes, in glaciers, mountain lakes and mountain peaks "only Gods, spirits or fairies (*pari*) live". Life, even if warmth is not detectable, implies that energy is moving about, for the process of living, for growth and reproduction.

Other forces also govern the world: the most difficult to perceive was gravity that was guessed, so to speak, indirectly from its effects but could not be understood through the weight and the fall of objects; the effect of the pull of gravity or weight of missiles and weapons in general was practically considered for marksmanship (Vannucci, in press), for building chariots and habitations and for all matters of daily life. Astronomical phenomena also were observed, measured and predicted through their effects and by markings, rather than explained. Prediction was necessary for seasonal migrations and pastoral activities, such as taking the herds to the mountain pastures or sowing at the right time of the lunar and solar cycles for best results. The relationship between lunar cycles, day and night (circadian rhythms) and the physiological cycles of women and domestic animals were known and taken into account, even if not explained. Ancient man lived in an Euclydean world as we still do in practice; in our everyday life there is no relativity or fourth dimension. In an Euclydean world which is the simple geometrical reality readily perceived by man, natural phenomena are cyclic, except perhaps for ageing of biological systems, including ourselves, though at present scientific methods of rejuvenation of tissue cultures break even that aspect of the **arrow of time**. The *Veda* must be read as a function of an Euclydean system of natural philosophy since all the speculative hymns are within the constraints imposed by the perceptions of the sense organs.

Fire and heat in nature were and are still used to improve the taste and nutritive qualities of food by boiling and roasting. According to Vesci (1985) the art of cooking the sacrificial offerings endowes on them a divine status; Vesci goes as far as stating that cooking is the most important part of the preliminaries of any sacrifice in the *Vedic* tradition, a statement however, that could be questioned.

Food and medicines

About fermentation and the role of fermented food, much could be said. Fermentation is a process that both requires and produces a certain amount of heat. Man has always availed himself of the **purifying** or we could say **disinfecting** effects of the fermentation processes. Fermentation in essence is the creation of a culture medium by specifically chosen microorganisms such as only that species of microorganism will survive while competing microorganisms are eliminated because they cannot live and multiply in the new conditions created in the medium by the dominant species of microorganism. In preparing fermented foods, man aids the selected species of microorganism to eliminate all other competing ones by boiling or adding some substances, in order to have a **pure culture** as desired. Different varieties of curds, yoghurts, cheese, wines, vinegar, broth and other foods owe their different tastes and properties to the original culture medium (milk, broth, casein) and to the action of the specific microorganism introduced by man as *inoculum* that induces fermentation and creates a monospecific culture. Fermentation increases the nutritious value of foods by increasing their protein and vitamin content and it also increases the lifetime of foods because it avoids **spoiling** ; spoiling is due to contamination by other micro-organisms which was avoided by the fermentation process itself. Contamination gradually changes the properties of the culture; spoiled cultures are correctly called **impure.** People living close to nature are great experts at handling fermentation processes to produce drinks, foods and manure and to loosen vegetable fibers from the matrix to produce textiles.

In addition to working in the field as draft animals, cattle give dung used as fertilizer and to make fire, they give food as milk, meat and milk products; ancient *Vedic* man was not vegetarian. Milk is thickened and sterilized or purified by boiling and is further made into *paneer* and curds by fermentation and butter is clarified as *ghee.* All these items last longer and are concentrated food more easily transported by man as he follows his cattle that moves about from pasture to pasture and during the seasonal migrations uphill and back. The bull is powerful hence Indra and Agni are often likened to the bull; the cow who is generous is called mother and the giver of the five gifts, *pañcgavya*, or milk, *ghee*, curds, urine and dung. The high content in ammonia of fermented cow urine and dung is an efficient disinfectant of the house floor. To this day the life of many Indians is a real symbiosis with their cattle; these practices are ecologically wise in that they provide full recycling of organic matter and avoid waste. Nature must be treated parsimoniously.

Milk and milk products along with barley and wheat were the staple food of ancient Āryans, though a certain percentage of them, like any other group of human beings, does not produce *lactase* in their stomachs after ten or fifteen months of age. Lack of *lactase* makes it impossible to digest raw milk. Among Indo-Europeans there may be as many as 15% or up to 20% or more people who do not produce *lactase* in their stomachs, these are the persons who say that they do not like milk or that they cannot digest it, which is true. Fermented milk derivates in their various forms, have been pre-digested by the microorganisms that cause the fermentation, hence human have no difficulty in digesting the milk-derivates. Shepherds of every culture know this well and feed their cheese and yoghurt to children and elders, ailing persons and puerperal women. "By kine may we pass over ill conditioned misery; by barley. . .over hunger. . .(*AV.* VII; 50 (52), 7).

The use of barley or wheat as staple food is indicative of the geographical origin of the *Vedic* Āryans who must have been seasonal nomads even before embarking on the long distance displacements which they later undertook. Rice was introduced very late in their diet and it is extolled at many places, specially in *AV.* XI; 2. Rice is still now a luxury for mountain people, dietary habits change with time, as imposed upon people by ecological constraints of the new environment.

Agni and the Āśvins are requested to send abundant food from the east. This could be interpreted as follows: in their eastern migrations the Āryans entered some or many Hindukush-Karakorum-Himalayan valleys that run roughly N-S or NW-SE where their descendants still live.

These beautiful valleys must have been heavily forested, the people that they met must have had a more varied diet than they had. Hence the request: bring us "good food", meaning: varied food, rather than the monotonous diet they must have been used to, and teach us how to plant it, raise it, collect it and hunt it. The same applies to medicinal plants, hence the invocation to the Āśvins. It was recognized that much could be learned from the local people: "the little girl of the Kirāta, she digs a remedy. . . with golden shovel, upon the ridge of the mountain" is said in a hymn against snakes and their poison; it illustrates well the process of synchretism of cultures in the making (*AV.* X; 4, 14).

Fuel

In trying to follow the *Vedic* people in their migrations towards the Indian sub-continent, we do not only perceive changes in attitudes towards a changed environment, but also a change in habits. Different foods would have to be prepared in different manners hence the use of fire would have changed accordingly; mud ovens and *culās* of the types still in use, must have been invented early, they are cheap and can easily be abandoned without any loss before the next season or displacement; easily they are made anew and quickly when necessary; they cost next to nothing, though they are not very efficient. There would also have been increased use of fuelwood and correspondingly more cattle dung would be available as fertilizer as agricultural practices were intensified and less dung was used as fuel. The extensive forests provided all the fuel-wood needed but at the same time the newly-come would have learned and accepted from the local people the concept of **sacred forests** (Gadgill, 1992; Ramakrishnan, 1992) and would have seen the value of good management practices.

The smelting of iron ore and making alloys with other metals required the production of high temperatures and efficient forges had to be invented or copied from pre-existing expertise; the *lohārs* are nomadic to this day; the simple forges that they use at present may have already been in use thousands of years ago. Agni is mentioned in association with iron (*AV.* V; 27-28) since no extraction from the ore and no steel can be made without fire, needed also to fashion tools. Fire was used to reduce rocks to fragments, cold water may or may not have been poured on the heated rock to burst it open (*ṚV.* IV; 3, 11); and the help of fire for this purpose is also mentioned elsewhere (*ṚV.* II; 15); "may we flaming break the rock that contains the prize of the contest". What is meant here probably is the breaking up of copper or iron ore before smelting: ". . .our ancients. . .clearing the earth they disclosed the red. . ."; copper is red, but copper ore is shining yellow, as chalcopyrite or is dark green as malachite or others; limonite which is a poor grade iron ore is rusty red and good grade iron ore is shining red or dark steel colour as hematite and magnetite or shining yellow as pyrite.

Three main environmental novelties would have faced the migrant *Vedic* people as they came to Saptasindhu. The first would be the encounter with major perennial rivers; the second would be the high-rise dense forests and a much greater variety of plants and animals; the third, seasonality of rains, the monsoon. Adapting to these changes required innovative capacity and the will to settle in the newly found haven as well as the will to learn from the local people.

Rivers

The first major geographical-ecological novelty encountered was the perennial great rivers with abundant supply of running water available for as long as they would settle along the river banks and fields. This would have been a marvelous situation for people coming from dry climates and steppes, no wonder that the rivers are repeatedly eulogized and invoked. Water was hailed as the great cleanser; extensive and intensive agriculture became possible. The dialogue between *ṛṣi* Viśvāmitra and the rivers (*ṚV.* III; 33) now called Beas, formerly Vipāśa, and Sutlej, formerly Śutudrī, does not only illustrate the sentiment of marvel that the Āryans felt in the presence of the great rivers, it also shows the urge to gain knowledge through understanding what goes on in nature; in fact the water cycle is accurately described: evaporation, cloud formation, precipitation, percolation, formation of rivulets and rivers, then the ocean and evaporation again. The hymn also describes how a dam caused by a landslide that blocked the river flow was destroyed by Indra's thunderbolt. Again here the life-giving role of water and its importance for agriculture is made visible. Interestingly the names of trees, rivers and places have not changed much since *Vedic* times and they are mostly readily identified at present. The reason why it may be difficult to know for sure what the *soma* plant was, could be due to the fact that probably many plants were clustered under the single name *soma;* in fact at some places of the *Veda* the word is in the plural form. Some of these plants had a **milky sap**, in other words they were latex producing which indicates that they probably are members of the families Euphorbiaceae and Asclepiadaceae which are native of the sub-continent and the sap of which contains poisonous alkaloids, some of them with curative powers or psychotropic effects.

Forest

Changes related to new vegetation cover in the new habitat may already have been partly met by the *Vedic* man during the migrations eastwards. The major novelty would have been the tall forests and dense vegetation of the pre-Himalayas. Food habits would have been influenced by the changed conditions and the daily menu enriched. The importance of goats and sheep would grow as that of the horse and cattle would acquire a different connotation.

To the use of the *soma* plant, whatever it was, was added the use of a great variety of local species. Timber would be available to build larger and stronger chariots and carts and horses and bullocks would have to be bred stronger. Houses could be built with the help of timber for rafters, windows and doors, walls and roofs could be made of stones and slabs of slate. The change from the nomadic to a settled life brought with it changes and enrichment of the material culture that can be clearly read in the *Veda*, specially the *AV*. Still man had to learn the peculiarities of the new environment and how to use it best.

Thus the Āśvins are more and more frequently invoked and historical studies show that the pre-Āryan people had to be befriended to learn from them. Additional uses of fire were introduced in the culture of the new-comers, specially important was the use of plants and natural products new to them, hence the ritualistic preparations, including rice, that required special treatment and the help of fire to be processed.

Climate

The seasonality of rains, the regular swelling and rising and even flooding of the rivers, the vigorous monsoon, the alternation of dry and wet seasons required many types of adaptations. These may have implied the adoption of pre-existing techniques of water storage and irrigation and many others, as well as the adoption of local expertise including for the determination of calendar fixes such as solstices and equinoxes. Appreciation of the culture of the local people implied the acceptance and incorporation into the *Vedic* pantheon of the Gods and relative myths peculiar to the new environment. For instance, the Kailāśa-dwelling Lord Śiva is identified with the *Vedic* Rudra by a process of assimilation and synchretism due to expanded cultural needs. Tolerance of the newly-met culture was an imperative necessity for survival and development. All of the above implies careful observation and study of the new world facing the incoming migrant tribes, it also implies adapting physically and culturally to the new conditions, a process that would have taken many generations of continuous efforts and many wars that are mentioned with pride and that are reflected in the frequent prayers for victory over the Dāsa and others. Lord Agni is throughout the Hotṛ, messenger of the Gods and benevolent leader. Fire as one of the physical manifestations of Lord Agni would also be used in an increasing number of new rites; most important cultural enrichment may have been the improvement of the smelting processes and of the manufacture of iron and steel objects, as well as the improvement of medical practices.

As routines were established and methods perfected, Agni became almost a **matter of fact** God; in rites and rituals and in homes his place was assured and insured, thus Lord Agni ever-present to this day, always invoked and respected, becomes less and less mentioned directly; physical fire, however, with the development of new techniques and the refinement of old ones, becomes relatively more and more important, more frequently mentioned and used.

Time Measurement

The *Śruti* as revelation received by the ancient seers was transmitted orally generation after generation until when they were put into writing by later generations, the major credit for this immense work being given to Veda Vyāsa. Besides traditional spiritual knowledge, mythology and practical teachings were incorporated, including music. Religious and other ceremonial rituals and rites in general were regulated, but ancient learned men had their own way of expressing themselves before the art of writing was invented. One way of expressing themselves was by building structures with a meaning, besides using a variety of signs and symbols. In Part I, I briefly mentioned the large amount of such structures in Europe that are now considered to have been created and probably used for rituals and astronomical observations. Some of them, where organic remains were found, were dated to be as old as 5000 BP with confirmed dates of around 2000 BC of many of them. The megalithic cultures that flourished in India were studied from the point of view of burial rites, hero and ancestor worship and are of a much later period. In April 1987, H.S. Bisht and I found two megaliths with hemispherical cup-like excavations that could be post-holes or excavations for grinding, lying on two hill-tops of the Kumaon hills. They are aligned and oriented and they could have given astronomical fixes much more accurately than would have been necessary for common agricultural or pastoral practices. I believe that also rituals and ceremonies were performed at the site, since there are numerous indications in the *Veda* that hill-tops were used for ceremonial and religious purposes even in prehistoric times. A separate study of these megaliths and of their possible cultural significance and practical implications is under way. In the present context what matters is that such pre-historic structures are evidence of the scientific pursuits of the ancient seers for the unveiling of Ṛta and thereby understand the non-living and living environment in which we live. We regard these megaliths as instruments in a scientific sense of which ancient man availed himself; but their use and meaning are lost in time. The present-day myths associated with these megaliths suggest their pre-historic antiquity and are supposedly related to the exploits of Bhima.

Most authors consider the numerous megalithic structures studied by palaeo-archaeology in Britain and Brittany, France (Ruggles 1988), including the famous Stonehenge, to have been associated with religious ceremonies. While this is probably true, my question is whether scientific discovery, the study of natural phenomena and the building of instruments to better understand the phenomena is not a ceremony in itself that entails a series of ceremonial methodical acts, related to religion **because** they are a means to acquire knowledge. The pursuits of modern science are a rite that similarly entails a long preparatory phase of study and the ritualistic performance of certain acts in a pre-established regular and planned manner, as the scientific method requires and as every laboratory or field manual teaches. Many authros go a long way to enquire why pre-historic man would have gone through so much trouble as building the structures. My tentative answer is very simple: pre-historic man shared with modern scientists the inquisitive frame of mind that is typical of man and that has made man, **Man.** Man cannot have changed so much in the relatively small number of generations of the past 10,000 years or less. Having found out something **significant**, man would then succumb to the challenge of knowing more and more, and for the purpose of understanding more he would accept and face the great task of building impressive structures to enable him to learn more and better from mother Nature. The word **mother** is here extremely significant. Man the scientist and engineer is basically the same all over the world, then and now. The reason why it took so long for man to emerge from the early thinking stages to some 10,000 or 12,000 years ago when he took a big leap forward, may be the small number of human beings on Earth to begin with, hence the rarity of the advent of a genius. Furthermore there was an enormous difficulty in the transfer of knowledge that could be done **only** practically and through example, being therefore restricted in space and time. Finally language was not yet fully evolved and codified and abstract thoughts and ideas would not have been easily expressed or understood.

In India, the amount of effort and time that would have been necessary to transport uphill stones weighing several tons and the amount of precise knowledge needed to align and orient them accurately would have been matched only by the determination of somebody and the importance given to the study of natural cycles. The knowledge obtained would also have been used for practical purposes, such as the ritual preparation of medicines and the performance of religious practices that are powerful psychological tools that give hope to man and strengthen his will and self-reliance. The areas or levelled hilltops where the megaliths of Kumaon are placed most probably were sacred areas for the performance of *Vedic* rituals.

Astronomical observations are important for nomadic people living in open spaces without landmarks, load stone or compass and at times when there were no roads or road signs. Megalithic structures, however, would only be constructed when migrants decided to settle down for good. Hence they were built respectively by Celts or pre-Celtic people and *Vedic* or pre-*Vedic* people, though they have come down to us as having no apparent purpose. *AV.* IX and X indicate markings of the calendar and measurement of the year and seasons and the course of the sun is given as 360 days (*ṚV.* I; 155, 6 *et al.*). *ṚV.* I; 136; 137 give hints of calendar determinations, mainly for fixing solstices and equinoxes. All religious practices are influenced by local environmental factors and in turn they play a role in the ecological perception of the relationship between man and God. Religions that consider that all nature was created for the benefit of man and that man is the master of nature, are permissive of ruthless exploitation of natural resources; on the other hand "primitive" and nature sensitive cultures always have some tenet in their religion about the sanctity of natural wealth, for instance of water and of the merit of conserving water. All religions mark the key festivals and anniversaries to coincide with natural phenomena of the lunar and solar cycles. For the *Vedic* people the winter solstice or the spring equinox marked the beginning of the year; the Kalash to this day commemorate the winter solstice and the spring equinox with prolonged and meaningful ceremonies related to rebirth, the awakening of nature after the winter depression, with fecundity and purification rites after the old year is disposed of.

Following the observation of sun, moon, cardinal points and their movements, man observed the stars, the eclipses and other astronomical phenomena, all well documented and paraphrased into the myths of the *Veda*, of the Chinese ancient texts, of the Mediterranean, American and Mesopotamian and other people of the world. There are no humans, however primitive, that have no myths and legends, which is one aspect of Vāk.

In the *Veda* usually the observations are correct, though the interpretation and explanation given is not always of the type that nowadays we would consider to be scientifically correct and valid. Man resorts to myths as a substitute explanation of the direct cause-effect relationship that he has not yet understood fully. For instance, solar and lunar eclipses were correctly observed to be caused by the shadow of the earth or moon cast on the other celestial body, but the explanation given of the phenomenon, was mythological. Conclusions, however, are usually sufficiently correct to be used for practical purposes.

One of the greatest merits of the *Veda* is to have put on record in the oral and later in the written tradition, the understanding and knowledge to be transmitted to the next generations for the benefit of all. The *ṛṣis* and the ancestors are duly remembered with infinite gratitude.

Ecology and Biological Cycles

Ecology is the study of each individual's and each species' or association of species requirements and of their limits of tolerance in relation to environmental factors. Ecology is the study of the exchange of energy and matter between living beings and between living beings and the non-living elements of the environment. Individuals live, grow, multiply and die and the species continues to exist and evolve. The inter-relations among individuals and of species among themselves and with their environment are the core of ecological studies; hence the best definition of ecology is "the study of the distribution and abundance of species and individuals". The environment, including other species present, is evidently a decisive factor in determining the presence or absence and abundance of individuals and species.

In conclusion, the study of natural cycles of living and non-living things, the analysis and measurements of the multiple aspects of Ṛta in space and time, is the true meaning of the word Ecology: the "science of the house". House here means the living and non-living environemnt of which man is a part, whether he wants it or not.

Of all natural cycles the diurnal or circadian rhythm is the most readily perceived. It is the first one to which new-born babies coming from the uniformly dark and warm environment of the mother's womb must adapt. The baby's chemical circadian rhythms are primarily regulated before birth by the chemical rhythms of the mother's organism, but the final adaptation to the alternation of day and night of newborns may take days or even months to be completed. Children learn quickly that nights are not all alike, some are shorter and some longer, but they also learn that the second most spectacular heavenly body, the moon, waxes and wanes according to its own laws. The day-night cycle brought about by the most important heavenly body, the sun, is something to which children get attuned automatically, they learn to welcome the short nights of summer and dread the long cold winter nights. Cycles of summer and winter, rain and drought, cycles of the flowering and fruiting of plants, of shedding of leaves, and of animal reproduction, nesting of birds and emergence of butterflies, migration and hibernation of all kinds of animals from insects to mammals, are concrete manifestations of Ṛta into which children grow often without a **conscious** perception of their knowledge. If we look back at our own past or observe the development of our infants and children, it becomes obvious that after birth we are at first assimilated into a world, or I would call it a mathematical hipervolume of knowledge and only gradually, and almost painfully, we begin to acquire personal, internal conscious **knowledge.** While understanding is a marvellous experience, the conscious acquisition and storage of knowledge is a heavy burden. It is a burden because it increases the personal responsibility towards oneself and others.

Children notice and learn the events of nature without being taught, except by example, so did early man; some of them gifted with specially inquisitive minds must have started to observe closely, to make correlations among natural phenomena by taking measures in time and space, seeking precision and accuracy.

Most studies of the ecological aspects of nature did not leave a record in stone but rather in the language of the *Veda* and most of all in the oral tradition. They refer to the five traditional elements of eastern and western Indo-European tradition, earth, water, air, fire and void or *ākāśa*. These are represented by the five headed *Nāga* common in Hindu and Buddhist sculpture. It is significant, for instance, that the bronze image believed to be a replica of a very ancient and large, estimated to be 3m tall stone image of Agoreśvara in the temple dedicated to this deity at Ikkeri in Shimoga district of Kerala and built about 1566-1570 AD, stands with the right hand in *abhaya mudrā* and the left in the *varadā mudrā*, under the protective canopy of the five headed *Nāga*. Innumerable five headed *Nāga* are represented, but with time more heads were added as sculputre and carving became more elaborate.

Mother Earth, Pṛthvī, often identified with Aditi, the infinite, the "free from bonds", as well as with water, air, fire and *ākāśa* proceeds from the down-to-earth beginning of everyday reality, to the imponderable, present but intangible space or void, the existence of which is inferred by reason rather than observation.

Most of the ecological traits in the *ṚV* and the *AV*, as would be expected, refer to the Earth, its features and produce, while the central figure is of course, man. For man's survival there is a constant preoccupation with food and medicinal plants. Vegetarian habits came late and only in the wake of Buddhism, hunting must have been a common practice in ancient times yet there is little mention to this type of activity in the *ṚV* though cooking of meat and other parts specially of the body of goats, is referred to. Shepherds must have relied heavily on the winter crops of barley and wheat for their staple food.

Rice came into everyday life only about 2500 years ago. There is still much daily ado among the people of the Hindukush, Karakorum and Himalayan ranges with plants, specially grasses, trees, medicinal plants and grapes. Where the word *soma* is in the plural, SS and SV have translated the word as **medicinal plants** or **medicinal herbs**, whcih gives a better interpretation of the passages. The *soma yājña* was an important ritual later conducted preferably on hill tops, about which there are frequent references; there are remains of stone alignments or slabs on many hilltops in the Kumaon on which I have climbed, that are markings of some sort. Sacredness of mountains and hill-tops is frequently mentioned (ex: *ṚV*. V; 41, 42) and still observed at present. The rites probably evolved into the preparation of several different plant decoctions with different effects on the human body, inclusive curative of diseases. Preparations were always made with the help of fire and invocations to Agni, the Aśvins, Agniaśvins and other Gods.

The migrant tribes did not only find an abundance of animals to be hunted and plants to be used for different purposes, they would also have a different type of soil under their feet and on which to build their habitations; the air and water, the whole climate would be different and have different forces. Powerful natural forces would have to be appeased (*ṚV*. I;151-IV;20) hence the supplication to understand what is going on: "give mental power to him who sings the sacred song" (*RV*. I; 151, 2) while the same hymn can also be read as a good description of an earthquake. Trees, forests, rivers, winds would be the good providers but also a constant menance when unleashed in their fury. A whole hymn is dedicated to the lady or Goddess of the forest: Araṇyānī (*RV*. X; 146), "the mother of all forest beings who tills not but has stores of food". The hymn also shows that flora and fauna were studied carefully. To this day the deity of sacred groves are Goddesses; I know of only one exception which is the very ancient masculine Puttu in Tamil Nadu who is the presiding deity of a mangrove sacred forest. The Goddess of the Sunderbans in Bengal is called Bańodebī or Banobībī and is prayed equally by Hindus and Muslims who would never venture into the forest without invoking her protection against encounters with tigers and other mishaps; they also need her permission for hunting, fishing and collecting other bounties as honey, wax, fuel-wood, timber, medicinal herbs, tan-bark from her domain, the mangrove swamp.

The association of man with forests was an expansion of the field of studies of the immigrant *Vedic* people. As pastoral migrants and nomads, the *Vedic* people were exposed, until when they reached the Balkh area, to a restricted environmental variety, physical and biological; steppes and savannahs are monotonous and extensive and people's best friends were their domestic animals, intimate partners in good and bad times, companions in joy and adversity. Domestic animals participate in family affairs and in the festivals of the community during which they are decorated with genuine pleasure equally for the decorator as the decorated; they are the pride of all. The sacredness of the cow was an ecological imposition since in good times the cow was the giver of all, the female of the bull, mother of the bullock and she was the symbol of femininity and the generator of new lives, the cow was the symbol of the mother Goddess, almost as much mother as Earth, *Pṛthvī* herself; the cow naturally became the embodiment of all that is good in the female world and the giver of the basic needs for the life of the body, including warmth; indeed mother cow is generous in life and after life. Even in times of adversity the cow would not be sacrificed since she was the only hope for the future and for the production of new bounties. The greatest praise to Aditi, the mother of deities, unbounded, unfettered mother of infinity is "Aditi, the cow, the sinless, injure it not" (*ṚV.* VIII; 101, 15). The bull on the other hand was respected as the embodiment of strength and power.

The wonders of the new environment are frequently referred to: "the big rivers of Sapta Sindhu can be forded" (*ṚV.* X; 113, 10) but it is important to know the right time and places where and when they can be managed. Indra, the deified leader of the people, king of Gods, is extolled in ecological terms, among others in *ṚV.* X; 134. The gentle gesture common to all shepherds whose wards depend more on trees than grass for fodder is described: "as for a goat. . .O Maghavan, draw down the branch" thus allowing the animal to browse on the lowered branches of *Prosopis, Mimosa, Acacia* that grow in semi-arid regions where pastures are scarce and most of the food comes from trees. "Head of the world is Agni at night" was interpreted by scholars as Agni being the moon though what is meant here obviously are the home fire and the camp fire that are kept burning all night. New environments, specially in the mountains, are inhabited by ghosts, demons, fairies and wild animals that must be kept away by fire and light, while man and his domesticated animals sleep. As centuries roll by, carnivorous animals once very abundant, become rare, but ghosts, demons and fairies become more pressing as the human population density grow and more people invade their dominions. Man continues to apply more and more pressure on the environment at higher and higher altitudes, provided as he is with the powerful means of modern technologies. Mountains slopes are deforested, fairies and *yakṣas* lose their abodes and cry loudly in the unchecked winds, demons let loose avalanches and landslides come down on their own, with no vegetation cover to keep the loose boulders and soil in place. Downhill floods become more frequent, widespread and devastating. . .but that is a sad story of the present, when the conservation practices as expressed in poetic injunctions by the *ṛṣis*, are no longer heeded.

For the *Vedic* man the mountain valleys are perennially great: "untouched by time, never lacking green trees and vegetation, with their voices they have caused heaven and earth to hear" (*ṚV.* X; 94, 12); "from the mountain came the pressing stones" and probably other tools as well. Stones are respected as an important and useful element of earth and mountains are the maximum expression of immortality: the hardness and permanence of rocks. The high peaks looming in the distance inspire elevation towards God. Stones are spared even by Agni who "eats **everything** else "who in his wrath shaves a hill as a barber shaves a head" (*ṚV.* X; 142, 4). Agni's fury can be terrible (*AV.* VII; 70 (73), 4, 5), the "Rudra that is in fire". . ."to that Rudra, to Agni, be homage" (*AV.* VII; 87 (92), 1). Fire and smoke are also used to frighten away enemies (*AV.* VIII; 8, 2). Agni is present even in water as dynamic energy in water flow, rivers, mountain streams, ice and snow avalanches, torrential rains that could be harnessed for different purposes; Agni becomes the associate of water for the creation of plant and animal life (*ṚV.* X; 142, 3, 4); "Agni descending with water impregnates in the plants" (*ṚV.* X; 21, 8).

And the *Vedic* people decided to settle in the wonderfully promising land. Knowledge had to be sought and was obtained from the earlier inhabitants, the local people collectively called *Anārya* and the resources of their territories are designated "the treasure of the *Paṇis*" (*ṚV.* IX; 111, 2). Their knowledge is represented figuratively by the myth of the stolen cows hidden in the cave, kept in the darkness of ignorance and liberated by the light of kowledge who is Agni as Soma Pavamāna or alternately liberated by the war power of Indra, the Great Chief. Significantly all of *Maṇḍala* IX is dedicated to Soma Pavamāna though the hymns come from different *ṛṣis*. The name Soma Pavamāna can be understood in many ways, Agnisoma, spirit and body, theortical and practical knowledge. Pavamāna means "self purifying", since knowledge is always correcting itself and only Agni is repeatedly called "self-purifying".

Plants

Plants of the new environment gradually acquired meaning and value as knowledge of their use was worked out or otherwise received from the local people; thus began the fusion of cultures. A corpus of knowledge was formed on the use of natural resources for building, for making roofs and containers for grapes juice to make wine, for water and milk; irrigation ditches, hewed out of rock and soil, appropriate agricultural processes and the use of harder woods for the plough and the hoe, for rafters, windows, doors and chariots were developed or improved. Instructions are given for making the *araṇis*, with the softer wood of the *aśwattha* tree or *pīpal* (*Ficus religiosa*) that is energetically rotated to create heat by friction and sparks on the lower, harder wood of the *śami* (*Acacia suma*), the appropriate type of grass to collect the sparks is also mentioned. Plants are "those that grew in old times". . ."much earlier even than the Gods" (*ṚV.* X; 97, 1) and are "different from many different places". Incidentally, this expression "earlier than the Gods" fully recognizes and admits the fact that the idea or concept of God was born in man's mind at some stage **after** man had gained the power of discrimination and had studied nature.

There are also plants that have "departed far away", this probably refers to those that were typical of the places where man dwelt earlier and from where he had migrated to new places having a different flora. There are also "plants with flowers and without, with fruits and without"; "plants that (*soma* is here in the plural) give us different things" which, in this context, means different natural products, including what probably are intoxicating or psychedelic drugs; "be glad and joyful in the plants" (*R.V.* X; 97, 3) and "bring back thy very self, O man" (*l.c.* 4). It also indicates that mixtures of medicinal plant extracts were prepared (*l.c.* 14) because they "help and lend assistance to each other". In *āyurvedic* medicine, mixtures are used rather than purified extracts of single plants. The *ṚV* and the *AV* on the whole contain a good summary of scientific knowledge of plants and animals, the latter being divided in biped and quadruped.

Finally it is Agni who says "in plants and herbs, in all existing beings I have deposited the germ of increase. All progeny on earth have I engendered and sons in women who will be hereafter" (*ṚV.* X; 183, 3). Agni is the energy materialized at the inception of a new life when the fecundation of the plant or animal *ovum* by the sperm cell takes place, it is at that time that energy becomes operative in the embryo of every existence (*AV.* V; 25, 7), plant or animal.

Indra, the powerful, is addressed as *Sahastramuṣka*, the one having one thousand testicles, to express his valour and strength (*ṚV.* VI: 46, 3); the reading of the corresponding verse in the *Sāmaveda* is biologically a watered down expression: Indra is said to be *Sahastramanyo*, or having a thousand powers; this expression according to Ludwig is more aesthetic, but biologically speaking it is evidently and undoubtedly less powerful and less correct. *Sahasramuṣka* is an expression that would come naturally to the mind of pastoral people given to animal husbandry and living in daily contact with their animals. It is sad to note that the western translators failed to see the powerful vitality of the expression and shyed away from the naturalistically beautiful and scientifically correct expression to adopt a less faithful but more prudish word as "power" or "ardour" that only poorly correspond to the joyful, healthy and full of life expression used in the *ṚV*. Incidentally, ardour comes from the Latin verb *ardĕre*, which means to burn. *ṚV*. X; 85, 37 provides another example of a prudish translation that is in reality an impoverishment from the healthy vigour of the original text. Where Griffith sees "pleasurė", the Vedic poet sees a normal biological function charged with ecological importance. Two renderings of the line:

यस्यामुशन्तः प्रहराम शेपमू।

are: "loving whom we embrace" or "in whom originated by desire we may beget progeny, or: men sow seed" (where *sepah* is a synonym of penis, acc. SS and SV).

or ". . .and welcome all my love and mine embraces. . ." (acc. to Griffith).

The Indra-Agni relationship is interesting in that wherever the dual divinity is invoked, Agni is addressed with intellectual and spiritual epithets while Indra is eulogized in concrete biological and human terms. It is logical to eulogize the great chief and leader, fearful in battle, winner of wars, the **king** of the Gods, himself deified because of his greatness as a conqueror with the epithet "*suhastramuṣka*", or more often, the bull. If Max Muller and H. Oldenberg (repr. 1990, p. 305) are correct in their reading, there are other biologically interesting examples. Fire is compared to the erect male genital organ or *praja*'nana (*ṚV*. III, 29) and the *pramantha* used in the generation of fire has the shape of the male organ. The *pramantha* is the upper fire stick and the two *araṇis* are addressed as "*mahī viśpatnī*". Agni as energy that drives life is said to "stir the wearied man to pleasure and great joy what time the longing gifts approach him as he comes" . . .In other words Agni is also physiological potence and his power can revive old age and impotency.

Maṇḍala IX is entirely composed of hymns addressed to Pavamāna Soma or self purifying *soma*. It is not clear how and why plant juices could be self-purifying. However, on the basis of several references in these hymns and because Agni is self-purifying, self-generating and is the life energy in all living beings, the hymns make good sense if they are read as the close association of *soma* with Lord Agni. *Soma*, extracted with the help of "mountain given pressing stones" becomes divine though not a God **after** being purified by fire. *Soma* is made pure by Agni (*ṚV*. VI; 8, 1), probably by boiling and distillation. Nature reveals its secrets to the seeker of knowledge (*ṚV*. VII; 5) and Lord Agni is the formless deification of the concept of understanding. But a form must be given to the concept and *Soma* often appears to be the mediator or the one who **facilitates*** the acquisition of the process of understanding and the acquisition of knowledge. "Soma Pavamāna teaches the secret name of the Lord" (*ṚV*. IX; 95, 2). *Soma* appears to function as a wise elder "preeminent for wisdom, leader along the straight path. . .by thy guidance our forefathers dealt out among the Gods their share of treasure";. . ."praise loving lord of plants" (*ṚV*. IX) and . . . "the mortal man in thy friendship has delight. . .the mighty sage befriends" (*ṚV*. IX). Agnisoma "in cooperation have revealed the light of heaven" (*ṚV*. I; 93,5), and are "makers of the path" (*AV*. XVIII; 2, 53).

Whatever substance or plant *Soma* would have been originally, it was obviously a substance that exhilarated, inebriated, intoxicated and gave **enhanced powers** mentally, physically and perceptively, perhaps by extrasensory perception. May be the *soma* plant was just grapes out of which fermented wine was and is still ritually made and

---

* Facilitation is a physiological process by which the transmission of the nerve impulse from nerve ending to the effectors, for instance a muscle fibre, is facilitated or quickened and made possible by pre-conditioning the receptor to accept the stimulus and to react to it.

consumed during the winter solstice celebrations of the New Year of the Kalash and the Kafirs to this day. One is reminded of the few-days-old wine made by Ulysses to inebriate Polyphemus and thus make possible his escape and that of his companions from the cave where they were kept.

There is an unending literature on *Soma* that is not to be discussed here, it appears certain, however, that with time other plants were added to the original one; some of them could have been producers of psychotropic drugs that give altered or enhanced sense perception. The word "*soma*" appears to have become almost a synonym of medicinal plants in general, thus the original bio-socio-ecological meaning of the word **soma** would have been greatly amplified. Maybe the revelation that is the object of the *Śruti* was in part obtained during periods of exalted awareness, hence the **visions** of the seers. There are hints that states of supersensory perception were induced by "herbs" (*Ṛ̣V.* IX; 25). The human brain also normally produces opiates and other psychotropic drugs like endorphins, the production of which may increase during periods of deep meditation, thus increasing perception.

Medicinal plants extracts would be included in the collective term *Soma* since they are something that interfers with the normal functioning of the body and mind: "flow on to Indra, Soma Pavamāna carefully effused, let sickness stay afar together with her friends" (*Ṛ̣V.* IX; 85). However, already in the first *Maṇḍala* the realm of *soma* appears to be made of many elements and the word *soma* appears to be a collective term not restricted to a single species of plant. "The herbs that thou hast generated" (*Ṛ̣V.* I; 91) and **Agnisoma** "have freed man from the fetters" (*Ṛ̣V.* I; 93, 5). Fetters of fear, in other words, the knowledge acquired through **revelation** disperses fear, as much as faith also liberates man from fear of death and worldly evils. Fear is the most powerful agent of destruction of mental balance, fortunately Agnisoma "found the single light for many" (*ṚV*. I; 93, 4).

Soma "bestows heroic powers (*ṚV*. IX; 43, 6—ṚV. IX; 53, 3) and also has healing power (*ṚV*. IX; 79, 2—ṚV. IX; 91) but many other plants are also recognized as having medicinal or intoxicating properties: "the juices (plural) with a thousand powers" (*ṚV*. IX: 42). The Vasiṣṭha hymns of the seventh *Maṇḍala* make frequent reference to medicinal plants and to their preparation, important elements being the pressing stones, water, the woolen filter, fire as fire and fire as Lord Agni who is joy of living and is knower of the correct rites and technicalities needed for the preparation of medicines. "Aid the pious through heat to life with your (*Aśvins*) assistance" (*ṚV*. VII; 69, 4); "delight yourself. . .in plants and waters when the *ṛsis* give them and ye find they suit you" (*ṚV*. VII; 70, 4) which is a hymn to the Aśvins, the medical experts with whom Agni collaborates. "The juices (*soma* is not mentioned, but liquids or water) that exhilarate are mingled, injure us not, but come with happy fortune" (*ṚV*. VII; 73, 4). And to the Aśvins. . ."man's dear priest (Agni) is seated who brings sacrifice and offers worship" (*ṚV*. VII; 73, 2) and in another hymn also to the Ashvins (*ṚV*. VII; 71) Vasiṣṭha alludes to the preparation of medicines through fire, where Agni is present as fire technically and as God for the invocations; as God he performs the rituals of the sacrifice. **Sacrifice** retains in all the translations of the *Veda* into Indo-European language the original meaning of "making sacred" or "sanctified" or "sanctifying". **Sacrifice** literally means "to render" or "to make sacred". In the Roman Catholic liturgy, the Holy mass is called the "sacrifice of the Mass" and the Christian act of the Eucharisty is partaking of the sacredness of nature through the oblation made to God of wheat and wine which are vegetable food made sacred. Thus the scholastic theory of the transubstantiation of bread and wine into flesh and blood is biologically and ecologically correct.

Agni is intimately related to all other elements that he

sanctifies by his own sacrality. Drinking of *Soma* excites and "gives direct access to the Gods" (*RV.* IX; 44) and *Soma* as the moon (Indu) has something to do also with mental disturbances, as is also expressed by the word "lunatics".

Āyurvedic medicines are often mixed with food, milk or honey to make them palatable or more efficient or both. Soma Pavamānah is described as of "many colours" and as coming from different places, including mountains. In fact also different parts of a plant have different colours and produce or store in their cells different chemical compounds that have different pharmacological properties and medicinal uses. "Coming from different places" could mean different species of plants that grow in different places; "Soma Pavamāna, a doer of many acts" is "dwelling in wood as Varuṇa in rivers" (*Ṛ̣V.* IX; 90, 2) and is also giver of enlightenment (*Ṛ̣V.* IX; 61, 16). Soma Pavamāna is gifted with a very important attribute, he "stirs the waves of sound, is by song a poet and a sage by wisdom" and "teaches the secret name" (*Ṛ̣V.* IX; 96, 17; 18). He gives the best gift, intelligence and wisdom, "let Indra. . .drink of the juice for wisdom" (*Ṛ̣V.* IX; 109, 2). An ancient Latin proverb brings to us a similar notion from the Romans: "*In vino veritas*" (truth in wine); it is recalled to this day at the holy communion of the Christian rites as it was at the Last Supper of Jesus Christ.

Great importance is given to the tree, the "Lord of the forest" . . ."whose praises never fail" (*Ṛ̣V.* IX; 12, 7); the tree is a symbol of life that always regenerates itself and is immortal, like the *aśwattha* on whom the Maruts rested, that germinates and grow in the air (above ground, on other trees or structures), which regenerates itself and from which the upper *araṇi* is made. The concept of sacred forest is very ancient and sacred trees or sacred forests are specially significant in the Hindu and Buddhist cultures. "Sacred forests" is undoubtedly a pre-Āryan concept and it was probably adopted by the immigrants because of its far-reaching significance and importance. Forest use and management would have been learned by the newcomers from the local inhabitants.

The perception of nature's cycles includes not only astronomical bodies, but plants and animals as well; "the soul of the plants, the fatness of the waters and let there be long life- ye Gods-to Agni" (*ṚV.* X; 51, 8) who is the energy that recycles everything; ecologically speaking it is the creation of life, its death and decomposition and re-creation of living matter through the all pervading energy of Lord Agni. Clouds are thus called the "son of fire" (*ṚV.* X; 31, 10-11) because of the heat that causes the evaporation of water, that after condensation into clouds pours down as rain.

Water, Fire and Life

Water is the great cleanser, second only to Agni, the Great Purifier; in contemporary terminology we would say that water cleanses and fire disinfects or sterilizes. "Waters have healing powers, they drive disease away; they have a balm for all, let them make medicines" (*ṚV.* X; 137;6,7); this reminds us of the fact that water is substance that has the greatest dissolving power.

An important role of all religions is their action on man's psyche; "the tongue that leads the voice precedes-then with our tenfold branching hands-with these two chasers of disease-we stroke thee with a gentle touch" (*ṚV.* X; 137, 7). Invocations, rituals, exhortations and magic power of chanted *mantras* sooth and strengthen a vacillating mind, spirit and heart, before or during the process of treatment of the body. They strengthen the will to endure painful interventions to dress wounds or set broken bones or other affected parts of the body. This verse may also indicate the use of massage, as is common for many ailments, and bone or muscular strain; massage is a common practice and a real science in *Āyurveda.*

Fire as life energy (*ṚV.* III; 22), as the energy responsible for the dynamics of nature or, in ecological terms, the agent of the continuous exchange of energy and matter and constant evolution and development of everything uninterruptedly, fire as Agni is the ecological factor causative of movement at all levels, from the molecular to the astronomical. Without movement, there is no life. Scientifically we measure heat in terms of temperature and we admit total absence of movement only at Absolute Zero temperature or minus 273^0C. The energy implied in the notion of movement is called in the *Veda*: Fire and heat or warmth, that "pervades the whole universe and being manifested as lofty *Viṣṇu* preserves life" (*ṚV.* X; 1, 3) while Agni is the "augmenter of food (*l.c.*; 4); in other words Agni as energy or heat is the cause of growth of living beings even those that are apparently cold, or that have the same temperature as the environment, which are all living beings except birds, mammals and some chemiosynthesizer bacteria and blue-green algae.

Within the laws of *Ṛta* it is "Agni who knows, who fixes rites and seasons" (*ṚV.* X; 2, 3). And again it is "Agni who corrects our faults and failings" (*ṚV.* X; 2, 4) and reminds "man weak of mind" to perform the sacred acts at the right season invoking each God at the proper time and in the proper manner (*ṚV.* X; 2, 4-5). In other words, it is the knowledge and wisdom of Agni which guide agricultural practices, including irrigation, ploughing, sowing, harvesting and post-harvest treatment of the earth's produce; he guides hunting and harvesting of wood, herbs and timber; he guides the mind and every act of everyday life. As the one who sets the seasons, the calendar, he is also the one who knows when the herds should be taken to the high pastures and when they should return. All acts have to be undertaken after appropriate propitiatory rites are performed to attune man with *Ṛta*, according to Lord Agni's instructions, since "he knows, he dwells in Ṛta". Cooperation of the gods has to be secured and man's confidence in himself has to be strengthened to generate determination, discrimination and steadfast appropriate action throughout his lifetime. Man's actions are regulated by the law, the ecological principles applied "with wisdom" to the various environments in which he lives.

Agricultural, pastoral and general living practices would have been different in the various places occupied by the migrant Indo-European tribes, but social and family rules and habits tend to be conservative. In fact most ancient habits concerning intimate family life have remained unchanged to this day in orthodox families, specially the habits relating to the impurity of women at certain days of the menstrual cycle and after delivery when she is specially vulnerable to infectious diseases and other ailments; the traditional care of pregnant women and newly born babies is biologically wholesome; mother and child are specially treated with appropriate massage, food, anointing, fumigation specially during the peri-natal period. For instance a decoction made from *neem* leaves (*Azadirachta* or *Melia indica*) is used for vaginal ablutions as disinfectant after delivery and in the cure of minor venereal diseases; where the neem is not available, decoctions of *Salix* (willow) bark and leaves is used in the same manner. Fumigation of women after delivery and making the puerperal woman squat over fume-generating burning plants "purifies" practically by disinfection of parts of the body at a time when those organs are in need of special attention and they create peace of mind and a feeling of well-being through its symbolic power.

Ancient man learned the laws of nature empirically, from experience and practice which, allied to his power of discrimination enabled him to improve steadily on his interaction with nature without transgression but according to ecological principles. Sacrifices and rites, attuned to Ṛta, acquire ecological validity.

Much of the *AV* is directly related to the use of many different plants, some of which still bear the same name and use. Important among the plants was the *kawa*, the "all-curing plant" that continues to be used in *Āyurveda*, together with other closely related species as ingredient in a large number of different preparations. But one is repeatedly reminded that all acts of preparation are necessarily ritualistic and that Agni, giver of force, power and strength is present at all *Vedic* rites (*AV.* II; 17, 1-3) because Agni "understands immortality " (*AV.* V; 30, 14); many rites are performed for practical purposes only, but all entail elaborate rituals and require the participation of Gods. Agni's blessings are invoked for success, for wealth and prosperity, for progeny, to conjure diseases, against sorcerers, demons, enemies, *bhūts*. *Bhūtā* is a word usually translated as ghosts and goblins though it is used in the *ṚV* and the *AV* for all "beings" and *bhūtā* is the spirit of each being (*AV.* I; 32, 1, 2). In *AV.* V; 24, 2 Agni animates even *bhūta*, he is invoked against witchcraft, savages, thieves and niggards, against snake poison (*AV.* X; 4), and he is also invoked to burn down the enemy and consign falsehood to the flames, he is even invoked against bad dreams (*AV.* XVI; 6, 9);. he is finally called upon to make imprecations of all sorts (*AV.* X; 5). Agni puts everything right with energy as the great purifier he is. The third of he five ecological elements, *mahābhūta* fire, is thus the link between the material and the immaterial, between earth and water on one side and air and void on the other.

The *Vedic* culture was pragmatic and could not have been otherwise, the philosohpical speculations of the Sages pursued at the same time two main streams, the abstract or speculative, and the pragmatic or the applications of knowledge and understanding to everyday life. "The purifying one (Agni as energy) entered into the green ones (plants)" (*AV.* X; 8, 40); the *AV* discriminates and shows the use of many more plants and animals than the *ṚV* does, and describes some technicalities.

The importance of the *pīpal* was mentioned above, it symbolizes life and recycling of life and immortality through its ever expanding capacity of putting forth new roots that eventually coalesce into a trunk. The ever fluttering leaves of the *pīpal* is evidence that the Maruts play with them and that at times they rest there; the *pīpal* is usually evergreen and lives for centuries, almost idefinitely and is considered to be masculine because it grows as an epiphyte on other trees, preferably the *śami*, considered to be feminine. The name of specific medicinal plants is sometimes kept secret, but several diseases and their remedies are described in detail and unmistakable terms, for instance malaria (*AV.* I; 25, 4—*AV.* V; 22, 10); spotted fevers, hepatitis or jaundice, leprosy, epilepsy, parasitic animals, specially worms (*AV.* II; 32, 3) and all minor common dieases. Even plants to promote hair-growth and virility are mentioned in *AV.* IV; 4. The plant against fever is said to be born on mountains and to be a companion of **soma** (*AV.* V; 4), it probably is the willow from which salicilic acid, the basic component of aspirin, is extracted.

Plants are also used in practices of symapthetic magic, against witchcraft, against evil spirits and super-human foes and exorcisms were often performed by fumigation. Other useful plants are specifically mentioned, like sugar cane (*AV.* I; 34, 5) grapes or the hemp and some others as yet scientifically unidentified. The forest tree is repeatedly eulogized for his heroic power or the one who relieves from the demon.

Fire, the third of the five basic elements is associated to earth where plants grow, to water and air that are needed for the preparation of extracts by boiling, by distillation, by fumigation and by mixing with cooked food. Agni is described at times as "flesh eating" (*AV.* XII; 2, 42) of which two readings can be made: fire destroys fever by the preparation of medicines and second the dead body is cremated by fire and sublimated into nothingness. The second reading is obvious in those hymns where fire is mentioned in the context of the performance of funeral rites, while the first reading seems to be intended when dealing with aspects of cures of fevers in general; but agni as fever destroys the flesh.

Besides the invigorating effect of fire, water is indispensable in the preparation of medicines, none however would be effective without the power of *mantras* recited or chanted by the sage and his disciples during the preparation process. Finally, since spiritually and ecologically everything is related to everything else, several other Gods are also invoked, most important besides Agni, is Savitṛ that in my opinion is the personification of the abstract idea of Life. Saraswati or knowledge and Brahmaṇaspati (*AV*. IV; 6), the Lord of Prayers are also present in the preparation of remedies and medicines as well as the Aśvins.

Agni and water are essentials of life "waters are possessed of glory in the plants". Agni is the divine perceived but not fully understood **life-process** that cannot take place in the absence of water. Life or the essence of life was and continues to be mysterious and unlike the global water cycle that is factually, figuratively, symbolically and pracitcally described in the *ṚV*, the *SV* and the *AV*, life and the absence of life are not often spoken of openly. The termination of life or death, on the contrary is often spoken of and discussed in its many aspects. It is recalled with respect and in a manner comparable to the manner in which all manifestations of life are discussed, because death is the beginning of a new life. When does a circle start or finish? All the points that make a circle are sacred, so even are the different stages of the cycle of life and death, not only because they are expressions and manifestations of Lord Agni, but perhaps as much because they are mysterious and remain essentially unexplained. They are the unknown of the after death, hence faith, religion and the theory of *Karma* take over. The spiritual importance of Lord Agni who as fire sublimates everything, is clear, cremation of dead humans was a custom introduced late, the ancients most probably enterred their beloved ones after death; as the Kalash,

Kafirs and Dards still do; also *sādhus* and *sannyāsīs* even nowadays are not cremated. Cremation is not ecologically wise in an overpopulated world though it is advisable from a sanitary and hygienic point of view. At a spiritual level cremation means liberating the soul from the body fetters, it means sublimating the material part of the individual, it means making the spirit immortal and identified with Agni. Cremation immortalizes the deceased precisely because it leaves no or very little material remains that are usually consigned to running waters, so that the cycle may be perpetually renewed materially as it is spiritually.

In fact it is Agni who "guards our bodies with ever watchful care" (*ṚV.* X; 4, 7) also because body temperature is an indicator of health or disease. Agni is eulogized as physician and maker of remedies (*AV.* V; 29, 1). In his mysterious interactions with life Lord Agni has also much to say about the functioning of the mind; "free the man possessed by a demon (*ṚV.* X; 161, 162) and Agni is also requested to cure tuberculosis or consumption of the body and many other bodily disorders, including leprosy and infectious maladies. Good health is taken as the normal condition, loss of health is loss of the norm of the law and therefore disease is often taken to be the result of a curse.

Agni and water are givers and sustainers of life, they are "affectionate mothers. . .givers of all, givers of life" (*ṚV.* IX; 2), they have healing power (*ṚV.* X; 137), but "most excellent are waters. . .most excellent is Agni. . .praised is the lightning cloud" (*ṚV.* I; 161). It is also made clear that the whole is the "most excellent", that each part plays its role excellently and that the whole cannot function without the adequate function of the parts, in this case, fire and water. The whole, whatever it is, cannot function if *Ṛta* is not observed and followed obediently. Soul and body are the Fire and Water pair that enables life to go on (*ṚV.* X; 11, 9) and Varuṇa's magic "puts fire in the water" (*ṚV.* V; 85, 2) figuratively as well as physically as body heat or swamp fire. Incidentally, this sentence was often misinterpreted and at best considered to be foolish, while it clearly refers both to natural phenomena and to the **fire of life** or life energy that cannot express itself in the absence of water. Common fire is the next indispensable ingredient for a better living process: "Agni accept with flames" (*AV.* V; 29, 15).

As mentioned, the global water cycle was fully understood by the ancient *ṛṣis*, beginning with the endless flow of water into the sea that remains unaltered; then "water vapour ascends to heaven" (*ṚV.* VIII; 41, 6-8; 42) and "Varuṇa is the large hidden ocean who envelopes the world with his robe (of humidity)". The burst of the monsoon must have been a striking sight, with the percolation of water in the soil and the emergence of springs and sources of water not available in the plains. The frequent lightning and copious rain following days on end of dense cloud cover would naturally have been linked to mighty Indra, whose thunderbolt is light and thunder in one. In fact there are good descriptions of the monsoon in hymns to Indra.

Clouds are important for people spending most of their time in the open; *Parjanya*, the cloud, is often remembered and one simile is biologically very expressive. He is requested to drop on earth the semen (*pānī* or water is a word that can be used, as well as *bīj* or seed, as a synonym of semen) of father sky that generates fecundity in mother earth (*RV.* VII; 101, 102). The famous and usually misinterpreted and misunderstood hymn to the frogs (*ṚV.* VII; 103) is a marvellous piece of good biological and ecological insight. Vasiṣṭha shows accurate knowledge of climatic and meteorological events and in this particular case also of the biological and seasonal cycles of different species of frogs, each with its own peculiar mating call. The "unexplained" verse: "cow bellow and goat bleat. . .and green and spotty (frogs) have vouchsafed us treasure. The frogs that give us cows in hundreds lengthen our life in their most fertilizing season" is a poetic rendering of a correct observation: good rains breed frogs in plenty since the tadpoles feed and grow in waters, and they are of different species, the green the spotty. "O Varuṇa. . .let the frogs. . .croak along the water courses" (*AV.* IV; 15, 12-14). The hymn also contains a good description of the female frog spreading out the four legs for reproduction; this is the posture she assumes to receive the male's amplexus. Good monsoon rains breed many frogs as well as good pastures and contented grazers such as cattle, horses, goats and sheep that will be healthy and productive. This is how plenty of frogs are a good omen for the next season. The hymn is very clear and not at all obscure, as most commentators would have. Un-necessary far-fetched interpretations are only secondary possible meanings if at all. Frogs are also a symbol of fecundity in pre-Colombian American cultures, specially in Central and South America and they also play a significant role in the shamanistic practices of the Turkish people of the Altai region.

The most important bio-physico-ecological aspect of water is that it is indispensable for Agni as the fire of life to become operative. Agni born in heaven is at home in water, he forces himself in water and becomes wood, cattle and their produce. The frequent appellation of fire as the grandson of water is not only due to the fact that living matter, twigs and wood included, cannot be produced without water, but also to the fact that fire or heat energy is necessary for the current biochemical reactions of the life processes to take place. The appellation of fire as Tanūnapāt, or son of himself, indicates that energy generates energy, that heat and food generate life-energy in the same manner as movement generates heat. Mental processes such as thought, meditation, concentration or *tapas* also cause an increase in body temperature.

Water, however, is not only the chemical compound H_2O that we call water. Even in contemporary parlance **water** is a synonym of any **liquid**, inclusive of medicines or semen. Soma Pavamāna is often described as flowing like a liquid or fluid (*Ṛ̣V.* IX; 30,1-*Ṛ̣V.* IX, 44,4—*Ṛ̣V.* IX; 45, 6).

Although the *Veda* are revealed Truth and do not contain detailed descriptions of the processes or methods followed for the acquisition of knowledge, some technicalities find their way in a more or less disguised manner in many hymns. Formulae, recipies or techniques are only given symbolically and obviously could be understood only by the initiated disciples who would have had the benefit of observing the acts of the *guru*, of listening to his teachings and assisting him in the manufacture or preparation of whatever it was (*Ṛ̣V.* IX; 71 *et al.*) the master was doing. Instructions include what would appear to be magic incantations that may well be procedural formulae or recipies. The whole process was accomplished by chanting, the sound and rhythm vibrations being an important element for the success of the process by conferring power to the maker and the made. There is a meaning in every gesture, in every act and in every *mantra*. The meaning can be objective, intellectual for memorizing or spiritual to influence the psyche, the mind and the spirit. The power infused in the preparation of medicines is either transferred to the user or will destroy the cause of the illness; "may he enriched with prayers come soon and early" (*Ṛ̣V.* IX; 93, 5).

To quote only a few of the descriptive lines: "*Soma* roars as he passes through the tail of the sheep" (which is the woolen sieve) (*Ṛ̣V.* IX; 43, 5), In fact fermentation generates foam and sound (*Ṛ̣V.* IX: 75, 4) and the sieve was made with the wool of the sheep's tail. Many uses of *soma-* or many different products "may go many different paths" or may be used in many ways (*Ṛ̣V.* IX; 78, 2). The change in colour of *soma* during the various stages of preparation (*Ṛ̣V.* IX: 78, 96) may indicate the preparation of dye stuffs many of which are indicator dyes, meaning that they change colour according to the pH of the solution. A common example is the preparation of indigo, another is anthocyanine that turns from pink to red, blue, violet as the flower first blooms, becomes mature and finally senescent; during the life cycle of the flower the pH (acidity/alkalinity rate) and oxidation level of the cell sap is what change and are the cause of the colour change of the flower, of indigo or of other species. In the preparation of dyes, the sequence and tone of the colour change indicates the correct undertaking of the usually elaborate process.

Different local populations and tribes knew well the natural products of the land, there is a long list of natural chemical products including dyes, medicines, psychotropic drugs, poisons, pain killers; they knew the properties of food, fibers and building materials, but it is not my purpose to discuss this here. I would rather refer to the exhaustive publication on plant, animal and mineral products having medicinal properties included in the "*India Materia Medica*" by K.M. Nadkarni.

Throughout the *Veda* we sense a deep respect for life, the most mysterious of all magics under whatever form it appears. Life is an important **manifestation and expression** of God, second perhaps only to Agni "giver of life" (*AV.* II; 29): there would be no life without Agni. Thoroughout the history of mankind and perhaps the pre-history as well, man looks up and inward when he thinks spiritually and about God and all that is good. The flames of fire are always turned up, as are the branches of the trees. "*Vanaspate* may you whom the sharp hatchet has brought for great auspiciousness, grow up with hundred branches, with thousand branches may we raise to greatness" (*ṚV.* III; 8). Thus the practical *Vedic* people recognized the presence of God in everything in the universe, including in the tree that they were about to cut down and they used everything in the best way possible, after the offerings were made and sacrifices conveyed by fire, *Agni Jātavedah*, the all knower.

Everything is sacred by virtue of its own nature because

energy pervades everything, thus the lofty tree is worshipped as well as the humble grass on which we sit and the one that helps ignite the fire; all and each one play their role in the cosmic symphony. Artifacts made by man, from the humble pot to the mighty chariot or, later, the great temples as well as the idols that represent Gods and are a late development, have to be sanctified by appropriate sacrifices because they are a part of nature though only secondarily so and through man's hands and mind. Vital breath, *prāṇa*, has to be infused in the idols at the most crucial moment of the ritual which is the sacrificial process itself. Materials are taken from nature to make things but as a new form is created for a new purpose, life has to be transferred to the new object from the whole from which the material was taken. Polynesians, the great navigators, had a similar behaviour: they invoked the spirit of the tree about to be cut for building thier canoes, specially the large travelling double canoes, then they ritually transferred the spirit of the tree to the finished craft, requesting its good offices to direct them in their migrations across the ocean and before embarking in their long distance sea voyages. Appropriate sacrificial rites have to be performed to abide by *Ṛta*, it is dangerous to fall out of step or out of the appropriate vibrations of the laws that govern the universe.

An interesting aspect is how the presence of thieves is taken for granted and in a matter of course manner (*ṚV.* VII; 55, 3); maybe it is far-fetched to see an ecological wisdom in the way people living in remote areas consider thieves and brigands, for instance in Tibet. Ecologically, however, this reveals an acute observation of the facts of nature: predator species of animals such as carnivores are needed for the maintenance of the ecological balance of multi-species ecosystems; diseased, old, weak prey are weeded out by the carnivores, while the gifted animals who run faster or who find better ways to escape the predators or to find shelter or who know how to defend themselves will be spared and in turn will generate healthier and better equipped offsprings. Likewise robbers are to be found everywhere in all human communities, but they are specially institutionalized, so to say, in sparsely populated areas at the margin of the parts of the world inhabited by man. They occupy an ecological niche that needs to be filled, human brigands and *ṭhugs* take the place of animal carnivores, they do not eat flesh but take goods which are matter, and money which is energy and thus provide for the recycling of matter and energy of the human population in areas where there is very little local production. As visitors, invaders, traders or caravans pass by, they bring to deserts, both geographical and social deserts, the essentials for survival that are not produced locally, so they have to be traded or stolen, or both.

Life is difficult to define, as a concept or idea, but is experienced as a *status quo*. Life is felt, is recognized, is lived and perceived when it leaves the body and stops being, but not when it begins. An hibernating bear, a leaf, an egg-cell fertilized or not, curds, any plant or animal are all living beings even if life is not readily discernible from movement or heat; these creatures will be seen to be alive when the bear warms up and moves out of its winter torpor or while the leaf remains green, the egg-cell of plants and animals divides and grows into an embryo. Plants and animals grow and reproduce while life is present, but when it stops, green plants become yellowish and putrefaction sets in dead plants and animals alike. Life is mysterious and all Gods are involved in the life processes directly or indirectly; if they were not, they, the Gods, would not exist. The greatness and mysterious aspects of life are inherent in Brahmā, whose consort Saraswatī is knowledge because there can be no knowledge without life, or as Descartes put it: "*Cogito, ergo sum*". Agni is vital energy, Savitṛ is the idea and abstract concept of life itself, Savitṛ is the glory of living, of the life-giving Sun, he is the Hope that daily rises anew after the fears of the dark night. I believe that the idea and concept of life was personified in Lord Savitṛ, and on this I will return later on.

Soma exercises power over the human body and mind. *Soma* **liberates** the instinctive and intuitive perceptions, a process **facilitated** by drinking special substances and *Soma* could represent the non-empirical, non-experimental, the unexplained or **revealed** aspect of the understanding of natural phenomena. The ancient *ṛṣis* must have realized that there are means of perception and communication other than those usually performed by human senses organs, as for instance telepathy. *Soma* was probably taken as one of the means to achieve supra-sensory perception. *Soma* is not deified but is glorified and exalted. Incidentally, the contemporary counterparts of telepathy are radio and television, these are means of communication the existence of which the ancients, innocent of the knowledge of electro-magnetic waves, sound transmission and electronic devices, might have guessed existed from observing the behaviour of animals. The orientation and migratory habits of animals who always return "home" safely, the orientation and communication among insects, as for instance the honey bees that were well known, the synchronized behaviour of herds of wild animals such as deer, were observed facts that must have puzzled the minds of the ancients. The mysterious means of communication and orientation of animals was translated in folklore as "animal speech" and animal wisdom; all the animals of the *Pañcatantra* speak and exhibit great knowledge, they are cunning, both good and bad. In fact elephants, whales, the latter unknown to the *Vedic* man, individuals of herds of hoofed animals communicate by ultrasounds to which man is deaf. Naturally the ancient perceived those manifestations of life and, unable to explain them otherwise, resorted to the *asu* or magic power of Varuṇa.

To this day there is no complete explanation of how migratory species of animals such as mammals, birds, reptiles, fishes, insects and many others, can navigate through air and sea over extensive geographical areas and return with pinpoint accuracy to their birth or nesting place. We know that some animals, for instance pigeons, perceive and are guided by electromagnetic waves; many mammals sense ultrasounds, such as whales, dolphins, elephants, bats and polarized light is perceived by insects and crustaceans and can be perceived by man only under forced special situations. It was Lazzaro Spallanzani of Galileo's school who made the first scientifically recorded experiments with bats showing that even when flying in total darkness they can avoid objects not thicker than a string; only about three centuries later it was proved that bats have a sort of radar-comparable navigational system for eco-location that enables them to fly safely in total darkness. Orientation by pheromones, by shock, or thermal waves or by humidity and barometric pressure sensors is common among flying and swimming animals that move in a three dimensional system or space; but though these devices were explained scientifically only as late as the XX century, they must have been guessed or perhaps even recognized by ancient man. Myths that refer to animal intelligence and "speech" are common all over the world and they are often similar among many people of very different cultures. The unexplained remained in the realm of *māyā* and the myth of the stolen cows besides probably reflecting one or more historical facts is also a symbol of the "hidden knowledge" that should be given to man for his benefit, since "our sires of old who knew the footsteps, found light and retrieved the cattle from the mountain" (*ṚV.* IX; 97, 39); "drinking of this that finds the light" (*ṚV.* IX; 108, 2). *Soma* gives mental power or understanding and strength to the body.

Air and *Ākāśa*

A few words about air. Air is recognized as breath of life, *prāṇa*, of all living beings, including fire that cannot exist without air (*ṚV.* X; 151, 4), also plants and animals cannot exist without air. The recurrent word *bhūta* indicates all beings and being alive means to have a spirit, therefore whether material or immaterial or just air, *bhūtas* have a real existence.

Air as wind can become dangerous (*ṚV.* VII; 89) and should be propitiated specially in the open plains and in the mountains, where wind can become highly destructive. Strong winds may uproot even the lords of the forest, they may contribute to trigger landslides and avalanches causing the formation of dams across rivers and impoundments of waters behind them, they extinguish home and camp fires and may incite forest and grassland fires to rage. Terrifying winds speak a language of their own understood only by ghosts, goblins, fairies and maybe a few among initiated men. But natural elements are ambivalent and *Vāyu* may also be beneficial as when he brings rains. Clearly, as later said the Buddha, the middle path is the best, not too much and not too little of everything. The peasants of my home country pray to St. Anthony when there is an excess of something, such as rains requested during spells of drought: "*troppa grazia Sant'Antonio*" (too much grace, O Sant Anthony), please do not give us too much of anything, even grace! In a general way, however, the cloud bearing winds, the Maruts, are welcome (*ṚV.* I; 167-169), which is natural in monsoonal areas where all nature thirstily awaits the rains after long months of dryness.

Air and water are means for the transmission of vibrations, visible or invisible, measurable or not, physical or mental (*ṚV.* I; 171) that produce sound, heat, motion and emotions. Other means of communication among men or between man and his environment depend on sound, music, rhythm, cacophonies that are all vibrations lying at the root of transmission of information and the offer of devotion. One of the great merits of the ancient sages is to have detected the visible as well as invisible forces, waves or vibrations that fill the air and the void: *ākāśa*. Sympathy or antipathy between persons arise frequently even at the first meeting, maybe an expression of such feelings cannot be expressed by words. We usually rightly say that domestic animals instinctively know if a newcomer is good or bad or indifferent , if it likes them or not and they behave accordingly. What is true of today's cows and dogs must have been true of their ancestors who lived with the *Vedic* man; animal's instincts rarely make mistakes.

The importance given to ॐ, which is the sound vibration par excellence that precedes all others in time and importance, was recognized to be so great, that it could not but set life in motion by putting man on the wave-length, so to speak, as the Gods.

Finally about the physiology of meditation. In recent years there has been much scientific research on different aspects of supernatural perception and achievements; this is no place to recapitulate the literature on the subject, suffice it to say that it is now well known that the amount of **dopamine** in the blood increases during motion and action. Increase of body heat during *tapas*, meditation and intensive mental activity, or just by the concentrated will to increase body temperature, may bring about altered states of perception that can be induced also by natural chemical compounds produced in the brain. These have effects comparable to those produced by some psychotropic drugs. Chemical substances closely related to opium and its derivatives, the so-called **brain opiates**; have properties similar to those of plant produced opium and also bring greater resistance to pain. *Yogic* methods achieve insensibility to pain, voluntary increase in body temperature, voluntary control of heart beat and many other feats, including expanded mental power.

From the observation of means of communication the pathways of which could not be detected directly by human sense organs, man sought to experience the same during episode of deep meditation or of trance (*Ṛ*V. IX; 18, 1- *Ṛ*V. IX; 20, 3). MitraVaruṇa's perception "with the eyes of the mind" (*Ṛ*V. VIII; 25, 9) is mentioned. It could be that the study of transmission of something intangible though an intangible medium produced the remarkable idea of *ākāśa* or void, that is comparable but not exactly the same as **ether** in ancient and medieval western philosophies. The concept of *ākāśa* in turn must have led to the remarkable concept of *śūnya* or void or zero which is a remarkable feat of abstract thinking. Zero has no intrinsic value but may acquire any value according to its position in any number, it is a direct logical consequence of the philosophical concept of *ākāśa*. It represents a remarkable leap forward from philosophical **theory** to philosophical **practice**, arithmetics in this case. The perception of *ākāśa* as one of the five fundamental elements of the universe is a much more subtle matter than the perception of earth, water, air or fire. It is a perception that goes beyond the understanding of life and is the recognisation of the much that we do not know about the processes of the universe.

Practicalities

Some further aspects of the application of theortical knowledge to practical aspects can be mentioned here. Health was perceived as the normal state, the maintenance of order in the body of man and animals as well as in the body of nature. Health is the law, ill-health is disruption of the laws that govern natural forces, due to ageing, a curse, witchcraft or poisoning. Health is physiological and ecological balance; disruption of order, *Ṛta*, causes biological and ecological imbalance, impoverishment or degradation of the patient, be it man, animal, plant or the environment. Degradation causes poor pastures, wear and disease in animals, decrease in productivity, all of them causes of migration in search of new pastures. When no direct relation of cause and effect was found either for the diagnosis or the cure, magic had to be resorted to; magic of course works very well for mental patients or hypochondriacs. Only healthy animals could be offered in animal sacrifices (*AV.* III; 29) to the Gods which is a hygienic practice of respect to the Gods and to the humans who would later consume the offering. A whole hymn (*AV.* III; 28) is dedicated to avert the ill-omen of twins born to cows. In rural Europe the habit persists of destroying twin calves of different sex because "they bring bad luck" and because in many places they are believed to be the offspring of a cow possessed by a demon. The biological basis of this story is that the male of a pair of calf twins of different sex is what in English is called a "free-martin"; it is a poorly defined male calf with all the characteristics of an intersex. At present we know the biological reason of the anomaly that is due to hormonal imbalance caused by the mixing of blood of the placentas of the female and male twin embryos. Obviously efficient herdsmen as the *Vedic* people were, they would have been horrified by such a freak, hence the label of "bad omen". Sacrifices had to be made to Lord Agni who sustains, maintains and controls *Ṛta*; logically, when disorder sets in, Agni is invoked to restore order. As Vd. Baghwan Dash (1971, p.8) says: ". . . all the active or productive aspects of *pariṇama* or evolution, at the macro-and micro-cosmic levels are due to *rājas* or energy". . ."*Agni*, according to the *Śaṁkhya* system". And Agni is the "kinetic factor at all levels of nature" (*l.c.* p. 22). *Āyurveda* also recognized different aspects of Agni, but to discuss this is beyond the scope of this study.

Interestingly Lord Savitṛ is invoked to keep away disease and in this context also the mountains, the pressing stones and the herbs who shed juices are taken together (*ṚV.* X; 100, 8); ecologically this makes sense, since the mountain gives the pressing stones and gives many of the plants used for different purposes, including the preservation of health.

Man interferes with the environment because basically man is not only a doer, but also a maker; the two combined produced man the tool-maker, the craftsman, the engineer, the artisan, the architect, the priest, the medical man. The pleasure that man derives in making things, the usefulness to everybody of what a few can make is stated clearly: "Atharvan first by sacrifice made man confident; for this skill the Bhṛgus were esteemed of all the Gods" (*ṚV.* X; 92, 10) and Agni Narāśamsa, the praise of man, Agni of all men, plays a decisive role (*l.c.*, 11). There is no doubt about the antiquity of fire, in fact fire has made man: **Man**. Man became what he is now by the use of fire for implementing different technologies.

In the *Vedic* mythology Tvaṣṭṛ represents the technologies themselves and the ability needed for the correct manufacture of object that are born from science and that are in turn generated by the understanding of Ṛta or the dynamics of natural and man-made systems. This is what techniques are.

The production of fire by attrition of the *araṇis* represents the sacredeness of the process and the need to follow the rules. Significantly Trita, a very ancient *Vedic* God, guides a priest, Agni, in the process (*ṚV.* X; 115). The uses of fire for different purposes are listed (*ṚV.* X; 115-122) where "these sages found what lay remote and hidden. . .from radiant Dhātar, Savitṛ, Viṣṇu, Agni, Bharadvāja who brought the *Bṛhat*". . ."They found with mental eyes the earliest Yajus, a pathway to the Gods, that had descended"; this is a good description of how knowledge and consequently the ability to make things was acquired. This includes all knowledge, spiritual and religious, factual and technical as well as scientifc, that all contributed to the development of man.

Fingers are poetically described as the "glittering maids" (*ṚV.* IX; 65, 1) not because as some translators have said, or perhaps not **only** because of gold rings, but principally because fingers do so much and so well. "See, all is done with these hands" is a common saying always expressed with great pride.

From the point of view of technical aspects of the *ṚV*, the IX *Maṇḍala* dedicated exclusively of Pavamāna Soma (AgniSoma) indicates the importance given to mental power as related to technical skills (*ṚV.* IX; 36, 3). Material and intellectual wealth make an indissoluble pair (*Ṛ*V. IX; 31, 1), as well as mental power and skill on one side and glory on the other, or intelligence and wealth (*Ṛ*V. IX: 32, 6). All the while Pavamāna Soma "Purifies. . .and sends the Gods who strengthen law" (*Ṛ*V. IX; 42, 5). Law in this context also means the rules of how to do things correctly to achieve an end.

At a later time Indra's power supports science and technology to ensure livelihood for man (*Ṛ*V. X; 112, 8): "in genuine wrath thou (Indra) looseneth the mountain (landslides) that opened the cave (where the *Paṇis* had hidden the stolen cattle) so that the *brahman* easily found the cattle". In other words, the king must be encouraged by men's prayers, instructed by the learned and protected by the Gods: significantly, the cattle had been stolen from the Lord of Prayer, Bṛhaspati. These are ecological laws applied to human society.

Finally, towards the end of the *ṚV* several verses refer to the means to preserve a functional society: in *Ṛ*V. X; 113, 5 the three powers who regulate society are linked together: Varuṇa's *māyā*, Mitra's law and Indra's sovereignty. A strong spirit of communality pervades the *ṚV*, the *SV* and the *AV*, hence the creation of *Saṅghas* and the recurrent recommendation of **harmony** among people (*AV*. VI; 74) and, I would add, between man and all nature. Food cooked by Agni Vaiśvānara ritualistically is recommended to unite people who have been separated by guilt or obligation (*AV*. VI; 119, 1, 2); Vaiśvānara, the Pacific will also ensure begetting offsprings.

The hymns attributed to *ṛṣi* Viśvāmitra, mostly in *Ṛ*V. III, are in general very practical, while the hymns by *ṛṣi* Vasiṣṭha in *Ṛ*V. VII tend to be spiritual and devotional, though they reflect a very good knowledge of nature and also teach many practical aspects of everyday living. The personalities of the *ṛṣi* can be outlined from their hymns.

Talking of technicalities I would end by recalling the

joyful mood of *Ṛ*V. VIII;1, that contains a detailed description of how *soma* is pressed to extract the juices, how these are filtered and later collected in three earthern pots, the *droṇakalaśa,* for the preparation of the beverage. The description in fact matches closely that by Robertson of the preparation of wine from grapes that the Kafirs make ritualistically and it also matches the description given by Lièvre and Loude for the preparation and consumption of grape wine by the Kalash. Fermentation is described happily as the voice of Soma and the description of how water oozes out of burning wet logs as "the water that rests in Agni" (*Ṛ*V. VIII; 102, 14). While the sound of Agni is joyous one can easily imagine the happy sounds coming from the hearth and the bonfires on the commons during festivals, specially the winter solstice series of rites and festivities.

The natural elements: fire, wind, water, air, seasons, birth, growth, old age and death of plants and animals, reproduction and reproductive cycles, floods, landslides, avalanches, earthquakes and intellect are all expression of the divine and they all fall within the realm of the eternal law and order or, alternatively, temporary disruption of *Ṛta,* or disorder. But on the whole, the hymns extol life and the joy of living, according to each one's *dharma* (*Ṛ*V. VIII; 35, 13), or duty.

In conclusion we may say that the sages were fully conscious of their duty to transfer the knowledge that they had acquired to other men; "the ancient names are brought to us by the songs of the poet who is a sage by wisdom, with the mind of a seer and the actions of a sage" (*Ṛ*V. IX; 96, 16-18). Thus, practicalities are not forgotten: "Purify the elixir through the filter"! (*Ṛ*V. IX; 97, 39-40); but this could also be read as: "Clarify your ideas through thinking, observing, and experimenting" so that you may live a healthy life.

"The wise man first comprehends the light of God in his heart,

then with three, the mind, the speech,
the deeds purifies the soul.
By his own nature he further makes himself most excellent
and contemplates on earth and heaven"

Ṛ̣V. III; 26, 8 (SS and SV transl.)

"With three purifying strainers he (Agni) purified the sun
with his heat the thought, discovering the light.
The mightiest treasure he produced by the powers of his own
nature,
then the looked over heaven and earth"

Ṛ̣V. III; 26, 8 (transl. Max Müller and Oldenberg).

5

Lord Agni and fire

> Agni revealer of Śruti to mortal men
> and giver of wisdom . . .
>
> *ṚV.* I; 31

> "O Agni, be unto us of easy approach
> as a father to his son:
> Stay with us for our sake"
>
> *ṚV.* I; 1, 9

AGNI as God is unborn (*ṚV.* I; 67, 3) or is "the first born", he is the institutor of worship (*ṚV.* IV; 6, 1); he is the one who knows mysteries and does wonders (*ṚV.* I; 77); he rises anew in ancient fuel, which can be read : he is always renewed in or by ancient wisdom (*ṚV.* X; 4, 5); and Agni himself declares his universality (*ṚV.* III; 26, 7. Lord Agni is not just common fire. Agni is the Lord of fire and he is "fireness" in fire (*ṚV.* I; 12, 6), "by Agni is Agni inflamed", or we would also say : Agni gives vitality to knowledge. Agni is the essence and divinity of fire, heat and energy. Significantly all the *maṇḍalas* of the *ṚV Saṁhitā*, except the VIII, open with a hymn to Agni and the *Saṁhitā* as a whole opens and closes with praises to Agni. All the hymns of *Maṇḍala* IX, including the first, are dedicated to Soma Pavamāna and can be read as addressed to special aspects of the dual divinity Agni-Soma.

Agni as common fire is born spontaneously or can be

kindled by man for instance by friction of the fire sticks, the *araṇis*. Agni as God is father of the Gods and as common fire he is their son (*ṚV*. I; 69, 1). As God Agni is observer of "heavenly laws of Gods and (as fire) of mortal men". He is giver of "wisdom that prolongs our lives" (*ṚV*. I;94) in time and quality. As God **and** common fire he is destroyer of evil (*ṚV*. II; 8,5) and "Agni is continuously to be found in fire" (*AV*. IV; 39, 9). "Agni bring Agni to the sacrifice" (*ṚV*. I; 44, 8) in other words: Agni fire (who art the means to an end) bring Agni God (who art the knowledge or truth, an end in itself), to the sacrifice (that men offer to Gods)".

The evolution of the concept of Agni as a non personalized God is clear in *ṚV*. IV; 1, 11 where he is said to have "first appeared as footless and headless . . . in the home of **Holy Order** . . . to fulfill the sacred law of worship (*ṚV*. IV; 1, 11-13). . . (*l.c.* 18). . . now all the Gods abide in their dwellings . . . brightly beaming Agni, the herald".

Agni as father and mother of mankind (*ṚV*. VI; 1, 5) belongs to all men as Agni Vaiśvānara; he is expressed as being one with Varuṇa and Mitra (*ṚV*. VI; 3, 1) and is requested to give the astronomical readings for the sacrifice (*ṚV*. VI; 2, 3), because he is invoker and inspirer of the universal rites (*ṚV*. VI; 1, 2). The Bharadwāja hymns repeatedly identify Agni with the Sun, which may indicate the habit of noting the position of celestial bodies to mark the calendar or otherwise make astronomical observations.

S. P. Saraswati and S. Vidyalankar, (1977, p. 423) derive the word Agni from अग = first and the root नी = to lead, thus they interpret the word *agni* to mean "the one who is involved first", the "foremost leader". They further recall that he is Vratapati (Lord of vows), that he makes everything part of himself and that he is also a drying agent. Dayananda (acc. to *SS*. and *SV*. , *l.c.*) has Agni as the Supreme Lord, venerable, adorable, omnipresent, respected by the learned and glorified by divine texts and sacred lore. Agni is the most brilliant (powerful) master of the intelligent mind. According to Sri Aurobindo (*l.c.*) Agni is Godwill and according to the same author (*l.c.*, p. 425), "*hotaram* is . . . the destroyer of *daityas*, a synonym of Agni Jātavedas". The usual rendering of the term *Jātavedas* is "omniscient". These meanings are all interrelated, because being omniscient, Agni Jātavedas has the wisdom to discriminate good from evil and as a *hotṛ* he has the power to destroy evil. Originally "Agni was the only immortal and later the other Gods received immortality from him".

ṚV. I; 1, begins with the invocation, laudation (Griffith), worship and magnification (Max Miiler) of Agni the adorable Lord, divine, high priest and Creator, keeper and sustainer of cosmic order, the one who works through and feeds eternal laws, who is fed with offerings of *ghee* and who sustains all that is light and luminous by night and mornings. He is immortal. He is the source of affluence, vitality and of interrelated knowledge and prosperity (*ṚV.* I; 147), he is the protector of the cosmic sacrifice of creation; he gives and accepts and he is indeed Truth itself. In terms of contemporary science we would say that Agni is energy **and** matter, because he is capable of interchanging the two.

"Agni is the first among immortals . . . he shall to mighty Aditi restore us" (*ṚV.* I; 24, 2). Agni is also *Viśvacarṣaṇir* (*ṚV.* I; 27. 9), well known to all. These seem to me, especially *ṚV.* I; 1, to be perfect descriptions of the root-idea and initial concept of God, and it well expresses the need and urge that man has to worship God. Agni thus becomes **the** all-pervading energy, or "soul" (*ṚV* I; 146, 1) "that is sunk in the lap of all that moves and moves not . . . and is ever youthfull" (*l.c.* 4). Agni is praised as the source of inspiration (*ṚV.* I; 22) "may he conduct the Gods hither", where Gods stands for all that is good, including knowledge. Agni is also requested to bring the spouses of the Gods, including Agnāyī (*ṚV.* I; 22, 9; *l.c.* 12), which is one of the few times when she is invoked. In fact Agni appears most often to be rather neutral than masculine or feminine, here it appears as if a gender had been given to Agni, as the husband of Agnāyī, almost in retrospect and out of necessity , because **all** the other Gods bring their spouses with them.

Agni's face is turned everywhere (*ṚV.* I; 97, 6), ancient, wise (*ṚV.* II, 3), the source of prosperity, material as food, intellectual as glory and spiritual as bliss. The sacrificial fire is nourished with *ghee* so that Lord Agni, therefore different from fire itself, may bring nourishment to man. Agni is the performer of sacrifice and the means that leads the seeker of knowledge or Truth. As the Hotṛ he is truth personified and "gloriously great". As Aṅgiras, Agni is the giver of all the good that is in him such as Truth which of course derives from understanding; he is dispeller of ignorance (*ṚV* . I; 74, 3) and is again worshipped as dispeller of night and darkness, synonyms of fear and ignorance. Agni is the one who brings light at the beginning of the day (Agni in the darkness of night or morning, and of the mind; *ṚV.* I; 1, 7). Knowledge is the means of our preservation (*ṚV.* II; 5, 1) and Agni is eulogized as the radiant, the sustainer of Ṛta and he is approached and worshipped in all humility with the most touching of prayers:

"Lord, be unto us of easy access as a father to his son.
Stay with us for our sake".

Lord Agni also plans for the future, by blessing "women with children" and "with motherly intelligence" (*ṚV.* X; 80, 1) and he instructs the bride to be vigilant over the domestic fire, the life and the human reproductive cycles. Lord Agni leads the bridal procession while Lord Savitṛ, the Vivifier, gives away the bride (*ṚV* . X; 85, 8-9). *ṚV* . X; 85, 40 is interesting in that it mentions Agni as the third husband of the bride, after Soma and Gandharva, while the fourth is the human husband, who is in turn, separately and previously prepared for married life. This could be interpreted thus: the future bride goes through a long period of preparation and training with the purpose of developing an excellent wife, mother and carrier of tradition to the future generation. As a child the girl child is wedded to "Soma" from whom she learns the home sciences, how to lead a wholesome and healthy physical life, how to prepare food and medicines and in fact the Aśvins come as wooers; the girl is taught how to bring up children who are inevitably always present in joint families (*ṚV*, X; 85, 42). To this day the importance of **didi** or elder sister in the inner family world is second only to that of the mother; Soma will also instruct the girl about family remedies and kitchen habits in general. Then Soma hands the young girl over to her second "husband" Gandharva from whom she will learn recitation of the oral tradition, myths, chanting and music. Finally the future bride is "handed over" to her third "husband", Lord Agni at the time when she is biologically ready to become a wife and mother. It is Agni who hands her over to her human husband after she has been fully educated and when she is apt to assume all the responsibilities of woman, the lady of the home. "Agni has given her to me for the sake of children and wealth" (*ṚV*. X; 85, 41). The bride fully concious of her role will bring happiness and prosperity also to "our bipeds and quadrupeds" (*ṚV*. X; 85, 43, 44) and "let her husband become the eleventh son" (*ṚV*. 85; 45). This indeed is a good poetical rendering of the four stages of womanhood: education and training, learning of tradition and rituals, physiological readiness for matrimony, consummation of the wedding. Even the Gods make oblations to Agni who is the moving energy of the physiological aspects of the life of animals and all mankind (*ṚV*. X; 88, 4, 9) and no marriage rites are complete without the circumambulation of the sacred fire. In *AV*. XIV; 1, 9 this is summarized thus: "Soma was the bride seeker, both Aśvins and wooers; Savitṛ gave Sūryā to her husband".

There is no better expression of the early formulation

of the concept of God and Agni as first God than the corner stone of the *ṚV*, the very first hymn. It clearly expresses the idea of God as a consequence of the human quest for intellectual, spiritual and practical knowledge. God is he who answers the many hows and whys of his child. God Agni is the fire that possesses the seeker after Truth, he is the inner spiritual fire (*ṚV*. I; 58, 1); he is the truth of everyday's life as well as **The Absolute Truth** (*ṚV*. I; 145, 5) and the law of nature of natural philosophy; Agni thus becomes the reality sought in religious and metaphysical speculations. As the human father teaches his son, God Agni at a higher level is invoked to enable the understanding and to disclose the truth that is above and beyond the common man's immediate grasp. The greatness and unrivaled beauty of this hymn lies exactly in rendering the agony of man in his fever to learn, to participate conciously in the drama of nature, which is the same as the fire of *tapas*. The *ṛṣi* says: "Take me - God Agni - take me as far as is can comprehend in spite of my human limitations". . . thou who liftest men to immortality" (*ṚV*. I; 31, 7).

Agni is one, yet many (*ṚV*, I; 127, 5) and "gives mental power to him who sings the sacred song" (*ṚV*. I; 151, 2); thus he is defined also as mental power (*ṚV*. I; 77, 3). Agni is the clearer of the path. Physically, fire clears the forest path and intellectually and spiritually he clears the path from ignorance (*ṚV*. I; 94). Fire as "bright red flames" gives strength to others, as for instance the Maruts (*ṚV*. I; 64, 7). Agni Vaiśvānara, as benefactor of all men (*ṚV*. I; 98) takes care of the whole universe; pervades all plants and vegetation and is Truth as the fire of understanding. In *ṚV*. I; 95, 7 he is said to "stretch his arms with might as Savitṛ", because Lord Agni as energy itself is also the energy of life.

Common fire

Different aspects of fires and their ecological roles are

described (*Ṛ̣V*, I; 95); a clear distinction is made between fires kindled by man and fires that "God have kindled" (*Ṛ̣V.* VII; 1, 22) that clearly refers to natural fires. Natural fires occur everywhere. In swamps, *apāmnapāt* could be swamp fire or methane produced by anaerobic fermentation of organic matter at the bottom of swampy waters. It ignites spontaneously upon reaching the surface of the water when bubbles come in contact with the oxygen in the air. Also phosphorescent organisms in the sea present in their billions may "put the sea on fire"; fire erupts spontaneously as bush and savannah or forest fires that are kindled by the decomposition of organic matter or are ignited by lightning. Two positive aspects of fire must have been recognized at about the same anthropo- geological time: fire for heating and drying and fire for cooking. Forest fires leave behind them roasted carcasses of animals and surely very early primitive man learned empirically that roasted meat is more digestible and nutritious than raw meat, one reason being that raw animal fat is hardly digestible while heated fat is more easily digested and absorbed; in addition ashes would have provided some salt that could otherwise have been deficient in the diet of very early continental man. It is clearly recognized (*Ṛ̣V*, VII; 7, 7-*Ṛ̣V.* VII; 42, 6-*AV.* IV; 35-*AV.* XII; 3, 24) that cooking increases the value of food, see also *Ṛ̣V.* IX; 66, 19 and *Ṛ̣V.* IX; 72, 9 where it is said that Agni is sent down to us with food for vigorous strength and "measure thou forth . . . splendid ample strengthening food, O Pavamāna". The use of fire for warmth and dryness brought with it the need to either store away fire as embers or to develop the ability to kindle fire anew. A small percentage of early men lived in natural caves where storing fire and keeping wood near it to dry it up for later use would have been an easy task once the observation of cause and effect had been achieved by some perceptive elements of the human community. Storing away twigs, dry grass and branches of trees for later use may have been one of the first methodical plans made by man.

Difficulties however were great for those ancient Hominids living in the open and worst of all for those living in wet climates or during the rainy season. Underground hearths, containers made of stone or hard wood to keep the embers may have been an early answer. Terracotta containers came late, when pottery was developed; "a receptacle for Agni" is mentioned in *ṚV.* I; 140, 1. Large number of perforated pots of different sizes and shapes are to be seen in the National Museum in Delhi coming from the very early strata of the Harappan and previous civilizations; similar ones are used for carrying and storing embers and light to this day.

How to make a fire was a crucial problem. Again we may imagine the most gifted among early men, much before the Harappans or any other ancient civilization, sitting and observing what nature does. Man would have tried to reproduce good Mother Earth's ways: either produce heat by letting grass fermentate in some corner of the cave or under some tree or fallen log or under a little stone hut made by man himself or he might otherwise have tried to make sparks to ignite some dry leaves or grasses. The techniques used would depend on the local climate and season. Flint stone lighters are widely used all over the world to this day in humid and wet places where matches, a recent invention, are useless or in remote areas where matches are a luxury quickly consumed and finished. Kindling a fire by attrition of two pieces of wood would have naturally followed; the *araṇis* are frequently mentioned directly or through metaphores throughtout the *Veda*, always with great respect and devotion. In conclusion, man learned that he could do what lightning does: he could ignite dry twigs and he could, by striking some stones against one another or by using meteoritic iron: "begat the fire between two stones", "Indra destroys the devil of ignorance which obstructs the path of virtuous thoughts as the lightning burns a tree" (*ṚV.* II; 12, 3 - *ṚV.* II; 14, 2). Kindling fire by rubbing dried sticks against one another became the ritualistic way of kindling fire. The surest practical way, however, would be to use intelligence to plan and not let the fire go out. To this day most rural people keep embers under cinders always burning. The sacrifical fire itself is ignited from the home fire.

The Ṛbhus said to be descendants of Aṅgiras are, in the *Vedic* context, the purşuers of knowledge who attained physical, mental and spiritual glory through their good deeds; they have a share in the sacrifies to the Gods (*ṚV.* I; 20, 8). Aṅgiras, one of the first *ṛṣis*, is said to have composed many hymns of the *ṚV*; he was ancient enough to be considered a *prajāpati.* His name may be derived from the same root as Agni and he is invoked as "God of fire" and "fond of sacrifice". Later myths have him born of the mouth of Brahmā, the Creator. **All** his wives or daughters have names of **ideas** or **qualities**. Aṅgiras is associated with Atharvan who is one of the most ancient *ṛṣis* who was also reputed to be able to make fire and offer sacrifices. The Āṅgirasa and the Atharvana are respectively descendents of Aṅgiras and of Atharvan and are frequently mentioned together as Atharva-aṅgirasa. They are experts in matters concerning ritualistic and technical use of fire. Later legends attributed to them supernatural powers and immense technical ability such as to rival the magical powers of an Atharvan called Dadhīca. The *vajra* was fashioned by Dadhīca and this legend, that would have a core of truth, obviously indicates the early use of fire to make metal instruments and weapons of **magical** quality and power.

From *ṚV.* X; 67, 2 it is clear that Āṅgirasa or fire priests were descendants of the brave and divine sages who first performed the sacred rites; in the hymns of praise they are requested to recite for posterity, reflecting correctly the "inner" recondite meaning and upholding the intelligent way of doing things.

The distinction of Agni as God, as well as the expression of God related to the intellectual and spiritual quests, and

Agni as common fire and the uses to which it can be put, is again expressed in the Vasiṣṭha hymns (*Maṇḍala* VIII). Many technicalities are taught by *ṛṣi* Vasiṣṭha, as for instance the purification of organic matter, presumably food, wood and animal hides by fumigation; Agni, the black bannered God, can be used to produce smoke, which is a common technique used for disinfection in many cultures. Fumigation is also part of the *Gṛha Praveśa*, and marriage rites. Among the Kalash, the Kafirs and the Dard of the Himalaya-Karakorum ranges and the South-Siberian Turks of the Altai mountains region, fumigation is used as a purifying agent for disinfection of homes and stables and as a routine practice, specially when they are visited by infectious epizootic diseases (Lièvre and Loude, 1990, pp. 50, 57, 72, 203). Fumigation by burning incense and branches of resinous trees and shrubs like the juniper and odoriferious roots and grasses, are also widely used to induce trances and for shamanistic and divination purposes; they are also a common practice in the rituals of many faiths. Burning sulphur, pyrethrum and its derivatives is commonly practiced to keep mosquitoes and other insects away, an important practice, since malaria may have been one of the single most important diseases of ancient times all over the world. Consuming food prepared through the good offices of fire and smoke is still a sacred, almost daily ritual practice. We do not owe our food only to mother Earth and it would not be available in good form without the action of Agni-God as life-processor and Agni-fire as food-processor. This is another example of the cooperation of several Gods for a single purpose; in fact, no ecological factor can act in isolation. Fermentation is a natural process used for purification of organic substances and one that requires the cooperation of several elements. The vats in which fermentation takes place are made of wood of the *pīpal* tree "made strong by iron, **ayas**, nails (*ṚV.* IX; 1, 2); iron here could mean metal in general. The fermentation process and all natural principles are mentioned as Soma Pavamāna, that is "everything and all Gods" (*ṚV.* IX; 5). *Soma*, the juices with a thousand powers (*ṚV.* IX; 42, 3) permit participation in cosmic affairs (*ṚV.* IX; 10, 8) and are the "divine elixirs of penetrating insight " (*ṚV.* IX 12, 7-9). Soma dwells in the hills (*ṚV.* IX; 18, 1) and Soma Pavamāna "graspest all things with thy mind and purifies thyself with thoughts . . . as such find us fame" (*ṚV.* IX; 20, 3). Agni and Soma are mind and thoughts (*ṚV.* X; 19), thoughts that can be put into practice as techniques for different purposes.

Common fire is often represented as Indra (*ṚV.* II; 11, 1 *et al.*) and eulogized with terms comparable to those attributed to Agni. The identification with Indra suggests that common fire and its many uses were quite distinct in the mind of the *ṛṣis* , from Agni God.

Among the ecological aspects of fire, we find Agni by day as the sun's energy and by night as the home fire and the campfire. May be it is also recognized that energy is always present even if not visible. Agni as creator and sustainer of life initiates the cycle with the photosynthetic process which is the production of organic matter in green plants thanks to water, mineral salts as nutrients and energy provided by sunlight; in several different contexts (*ṚV.* V; 44, 1) we are reminded that this is how life is created though the details of this magic could not be explained. Fire is represented as kindled by Tvaṣṭṛ's daughters as the ten fingers moving the *araṇis*. Agni also has many roles as sacrificial fire (*ṚV.* I; 95, 3), as astronomical fire, as the sun who sets the seasons; fire is latent in waters and water is indispensable for life, including for producing combustible material such as wood; Agni is swamp fire or methane and finally Agni is the energy that moves the water cycle (*l.c. 10*); fire is also used to burn away dead grassland to promote the sprouting of fresh grass and in a host of other ways. Agni as sun marks diurnal and annual cycles and early man was impressed by nature's recurrent cycles of time. Perhaps the only unidirectional aspect of time that he must have perceived was the process of ageing of all that lives. The Sun, the prime giver of life on Earth is an expression of Agni as energy and of Savitṛ as Life and Hope; the sun sends his vivifying rays downwards while fire on earth sends flames upwards; Agni is the go-between, the messenger between man and Gods, he sublimates the offerings made by man to the Gods and in return gives the imponderables: life, thought and energy. Nothing in this world would be possible without Agni: "O Agni, thou on whom all other fires depend" (*ṚV.* VIII; 19, 33) in accordance with Ṛta which in the present context is all the ecological principles on earth. Rituals and the proper pronunciation and intonation of the *mantras* are the religious expressions of the true knowledge perceived and transmitted by by the ancient *ṛṣis*.

The Aśvins are related to the physical life of the body as such they are closely associated both with Agni as God and agni as common fire; "let the fire divine bring oblations to us again" (*ṚV.* X; 24, 4, 5) that can also be read as an allusion to the preparation of decoctions, infusions and artifacts. Meat was usually consumed after boiling as is still the habit among the Kalash, and the Aśvins are requested to "churn out the fire" (*l.c.*; 4). We can easily visualize the *ṛṣis* and the "*ācāryas* who came after them, preparing food and decoctions ritually and studying their effects on man and animals . Purification and concentration of extracts "fifteen fold" to prepare "strong juices" by fire is indicated in *ṚV.* X; 27, 2. All of these are habitual processes practiced in *Āyurveda*. While the material preparations are done by fire, songs and incantations and therefore sound vibrations accompany the process (*ṚV.* VIII; 35). "The Aśvins coming from the east may bring plenty of food and medicines" (*ṚV.* VIII; 5, 31 *et al,*). The *Dāsas* or earlier inhabitants are specifically mentioned as knowing better food and medicinal plants of a rich flora and fauna previously unknown to the newcomers (*ṚV.* VIII; 9).

From a biological point of view man is a diurnal animal

guided by sight. Light is the main requirement for man's orientation; at night, fire is the natural substitute of the sun's rays. By day, light from the Sun, who is an expression of Agni enables man to go about his business; Agni is represented in the sky by Sūrya and is honoured as such (*ṚV.* X; 7, 1-3). Therefore Lord Agni is the one who illumines man by day and by night, physically and metaphysically as well as symbolically. Lord Agni enables man **to see**. Agni is implored to come out of darkness (*ṚV.* X; 1, 1) as the one who subdues the gloom of night (*ṚV.* I; 143, 3-*ṚV* . X; 1. 2) while "Soma teaches the secret names of the Lord".

Lord Agni-God

"O Agni. . . through all thy fiery glow purify this prayer of ours. . . cleanse us with thy own cleansing power, purify me on every side. . . Savitṛ God by both of these libations and purifying power (material and spiritual) purify me on everyside". Savitṛ appears here fully in his role of Vivifier, Impeller, the personification of Life and Hope, because through the heat and energy of fire he can maintain or restore health and life and he can give hope for the future. In conclusion it seems obvious to me that Lord Agni is the path to knowledge through understanding (*ṚV.* I; 72) that stimulates the need to perform noble deeds (*ṚV.* I; 79, 9); Agni is therefore both the means to perform rites and knowledge itself; Agni enables man to take the appropriate action deriving from knowledge. By meditation **vision** is enhanced: "Eyes that Soma gives" (*ṚV.* I; 139, 2),

Lord Agni in its fearful mood as anger and wrath and as destroyer of Rudra, received a different name: *Manyu* (*ṚV.* X; 83, 84) whose main duty is to destroy enemies.

The closing hymn of the *ṚV* is also dedicated to Agni (*ṚV.* X; 191) and shows a fully evolved concept of God. Lord Agni is now, in the present tense, the manifestation and trustee of the socio-political structure of the Āryans. The hope is expressed that man may have learned his lession and that human beings may be able to unite for the welfare of all. Agni, God, is the power needed to ensure that this will take place, he enjoins men to "assemble, speak together; let your minds be of one accord. As ancient Gods unanimously sit down to their appointed share". These terms are typically those of an evolved, well formed and mature religion that relfects a well structured society. Here Agni is the most ancient of the overlords, the wisest, the one who has the keys to the Absolute Truth as well as the keys of human truth. Agni has knowledge as well as power to impart that knowledge and to guide and direct those who should disseminate that knowledge. Agni the mighty one, is the one who obtains all that his friends and worshippers require and he is requested to bring to man **all Gods** as he is enkindled. Thus at the end of the *ṚV* we find Agni as a fully developed totipotent God, no longer just an idea of what was perceived and of the much more that is not yet understood by man. Agni is the one who can direct the thoughts of **all** those assembled for worship; he can make them of a common mind and purpose; Agni more than an individual God has become a social and political God as well, though he has retained the exclusive quality of being **The One** worshipped in the Assembly. Man may learn his lesson and may human beings be able to unite for the well-being of all.

Agni has grown from being **Knowledge**, to the personification of **Wisdom** itself. Agni is requested to "make us sinless in the sight of Aditi the original human mother, the one who cannot be divided, she who is free from bonds" (*ṚV*. IV; 12, 4) who is here invoked as the primeval giver of mental and spiritual humanhood. Agni is acclaimed as the guide, teacher and the one who helps men who have become "weak of mind, of feeble understanding". . . "then the prudent discerning Agni worships the Gods. . . in season" (*ṚV*. X; 2, 5). In other words, man recognizes his short-comings and to compensate for them he establishes rituals. Lord Agni illumines the path of eternity, thus satisfying man's desire for fame and for immortality, while

Prajāpati is the first human father, whose antiquity is lost in time, who is as ancient as the beginning of man and "the one who instituted the *Vedic* sacrifice".

"Agni understands immortality" (*AV* . V; 30, 14), the *ṛṣis* of course knew that immortality of the body is impossible, but they postulated that immortality of the spirit, which is the energy and heat of Lord Agni that resides in the living body, could be attained. In fact, through the enunciation of the *Śruti* the *ṛṣis* themselves attained immortality in the minds of men and in mythology as the constellation of the *Saptarṣi.* They taught men how to reach the goal of immortality by participating in the rites (*ṚV*, VIII; 44, 30) and praying to "prolong our lives" (*ṚV*, VIII; 48, 3) "we have drunk *soma* and become immortal; we have the light of God discovered". The powerful effect of sound vibrations is also taken to "give kinship with the Gods" (*ṚV*. IX; 10, 8); as well as the exhilarating power of *soma* and the enhancement of man through sound and thought. Rituals duly performed help to make man and all the nature of this world one with the universe, hence the conclusion that all is sacred. Would this be the great mystery of all mysteries? Would it be the beginning of esotericism which is at the root of the secrecy of religions that are entirely revealed, if at all, only to the selected few? (*ṚV*. VIII; 44, 28-*ṚV*. X 85, 3) . . ." and let this man, good Agni, be with thee the singer of the praise: be gracious, holy one, to him". In fact these particular hymns are a good example of how philosophy of science becomes religion through faith in God who is the only omniscient and omnipotent. Hence the need to perform rites and rituals, to utter prayers and to be initiated. To rituals and prayers always new details are added to improve on their efficiency (*ṚV*. VIII; 45, 1); "kindle the ritual fire to purify your heart". In the *Vālakhylya* hymn XI (*ṚV*. VIII; 59, 6) it is said that "God gave to the *ṛṣis*: revelation, thought and power of song".

The meaning of the first and last hymns of the *ṚV*

Saṁhitā (*Ṛ*V. I; 1-*Ṛ*V. X; 191) is direct, concise and clear. Though rendered with slightly different variants by different commentators, the basic meaning is the same in all of them. There is no misunderstanding about what the two *ṛṣis* of the hymns at subsequent times had in mind and what it was that they had seen with the eyes of their mind. In comparing the two hymns, the evolution of the basic concept of the primordial God Agni into a well established, powerful and generally acknowledged God at the end of the *Saṁhitā* leaves no margin to doubt. These two hymns are explicit and only one interpretation with no double meaning is possible. This is that at the beginning only Agni is worshipped, eulogized and prayed to with the full conciousness and firm belief in the power and greatness of the Idea of God who **Is** Truth and Knowledge: "Agni fill me with splendour" (*Ṛ*V. I; 23, 24). Later the concept of God is extended to other ideas and powers, but Agni remains omnipresent (*Ṛ*V. I; 27, 2). At the end of the *Ṛ*V , Agni is a social God with policy and political power.

The dual divinity Indra-Agni is worshipped as wealth, Indra, and as rays of wisdom, Agni, while the vital breaths are the *Śruti*. The physical counterpart of this allegory is the need of air for respiration (*Ṛ*V. I; 112, 4) and combustion, hence fire is described as the son of wind and of mother sticks, the *araṇis*. Accordingly, there are three types of fire rituals and oblations: (1) with food presentation and presence; (2) with *ghee*, for the sublimation of matter into spirit, and (3) for purification of dwellings and preparation of medicines for the body and the spirit. Agni is master of three worlds; physical, mental and spiritual (*Ṛ*V. I; 146, 1) and extends over time and space.

In the Christian culture, St. Francis of Assisi who also lived in close partnership with nature, in his "Chant of all Creatures" (*Cantico delle Creature*) sings:

"Praise to you, my Lord, for brother fire

> by which you illuminate the night and
> he is beautiful, jocund and strong".

To St. Francis fire is "joyful, beautiful and strong", in the *ṚV* fire is "vigorous and loud voiced". Throughout the Judaico—Christian, Graeco-Roman, Druidic, *Vedic* and Hindu-Buddhist tradition, fire trials were common as "divine judgements". At present fire walking is reputed to be safe only for the pure at heart. Fire has always been used to punish or to test virtue, to destroy enemies and give supernatural powers to such as can manipulate it, such as the alchemists of medieval and Renaissance Europe. Flames of fire are seen as a halo around Gods and Saints of all religions (Vannucci, Fire and Marksmanship, in press).

The dual divinity Indra-Agni appears to indicate the recognition that man is the sum total of physical power (Indra) and mental and spiritual power (Agni); a significant representation is "O resplendent Indra-Agni (Wealth and Power) I looked around longing for weal" (*ṚV.* I; 109, 1, 2 *et al,*).

The most common epithet by which Agni is greeted (*ṚV. I; 1 et. al.*) is "adorable", but others are also frequent (*ṚV.* I. ; 127, 1, 2) such as divine keeper of truth and cosmic laws, luminous, resplendent, shining, giver and acceptor". He is eulogized as messenger to the Gods and the one who brings the Gods to the sacrifice; he sits at every rite (*ṚV.* I; 14, 11) intermediate between men and God, he is the ancient (*RV.* II; 3, 1) and ever young.

Fire in agriculture and in everyday life as dispeller of humidity, fire as drier of green and wet firewood and animal hides and pelts, is as important as fire giver of light and heat and fire for cooking. This is manifest in both the *RV.* I; 140 and 141 where Agni the priest is lauded for his use in agriculture and even more so in the *AV* where it is obvious in the repeated importance given to the *araṇis* and to the dried grass that receives the first sparks. Technicalities about the use of fire are given in some detail among other places in *ṚV.* III; 29 and *AV.* VIII; 3; in the present context what should be pointed out is that the technicalities and practical aspects of dealing with and using fire are not to be confused with Lord Agni, though they come under the same name.

The repetition of the same adjectives for different Gods and the fact that most of them are in turn worshipped as "**the** Greatest", "**the** Mightiest, **the most** Wonderful, **most** Lovable, **most** dorable, **most** Luminous and so on has created confusion in the mind of translators who cannot see how each and every God can be **The** greatest. This is best explained by the reply given by a mother who was questioned about how could she truthfully and honestly write to each of her many sons: "My **dearest** one"?; the mother replied: "Each son is truly the dearest". This answer that all mothers will find natural and self-explanatory suggests the notion of a "collective Absolute", a syncretism of all and everything that is one of the infinite characteristics of the Godhead. However, the frequency of superlatives in the *Veda* has often been mentioned as lack of consistency, or lack of will, intention or competence of the seers to draw a clear picture of the *Vedic* religion; it has also been taken as a result of their inability to draw rational theories and to formulate a consistent religious doctrine. Ecologically speaking, however, since everything is related to everything else and since under different situations and at different times it is one or the other ecological factor that is the over-ruling or decisive one, it is as if a particular God or ecological factor is the one who does, i. e. the **factor** that is preponderant and is the regulatory element under the particular conditions and purpose being considered. It is ecologically correct to eulogize, address and consider as topmost different Gods under different contexts of place and time. Each God occupies a particular ecological niche, hence each factor or each God is addressed in turn as "**the**

greatest" because he is **de facto** the greatest, or the determining factor, or the most important factor at the particular time and place. It must be made clear that most Gods are ideas and factors, but not the natural phenomenon itself.

The complexities and interactions of nature's ecological principles is also rendered in the multiplicity of meanings of single *stotras*, which again is an ecologically valid approach because every natural phenomenon is acted upon and in turn reacts to many different environmental variables at the same time. Ecological factors (from the Latin verb: *facĕre*=to do, to make; factor=the one who does, who orders) are the elements or impinging forces. Indra at times cannot be easily differentiated from Agni (*ṚV.* I; 15, 3) and from the Sun, Sūrya by the epithets addressed to them but it is clear from the context of the hymns that they are different personalities, different principles, different Gods or deities. Thus most Gods of the *Vedic* man, as reflected in the *Veda*, mostly represent definite concepts rather than natural entities.

There is a great similiarity also in the approach to Agni and Indra. Indra is the king of the Gods and as Dandèkar has convincingly shown, he would have been a successful chief of the Āryan tribes in their eastern migrations and occupation of *Saptasindhu*; in fact the word **King** is more a human than a godly title. The deification of such a great leader is natural and God Indra has retained a number of purely human attributes and weaknesses. Leader of men and leader of Gods, conqueror of lands and people, giver of wealth, God Indra is one who is typically not the deification of a concept, but the deification of a great man who was giver of material benefits. Indra is also termed resplendent, noble, beneficial, glorious and of supreme intelligence, Indra brings light and form (*ṚV.* I; 6, 3) to men as well as material wealth through his ability and valour in conquest. The thunderbolt and the lightning powers of Indra are more related to what in French would be described as "*foudre de guerre*" (lightning war-power) and his attributes are more generally linked to battles, wars and conquests. He is the great Lord, but he is neither the primordial nor the supreme God, He is said to give light and life to the mind (*ṚV.* I; 6, 1) as the rays of dawn (*l.c.;* 3, 5), the one who brings wealth and is of adamantine justice (*ṚV.* I; 7, 2). But there is an important difference between Indra and Agni; Indra is **made** a God because of his great deeds, while Agni **is** God by virtue of his own nature, because he is understanding, truth, knowledge and the immortality to which man aspires. Agni resides in the human mind, as part of the cosmic energy. Indra (*ṚV.* I; 8) brings fame, victory and power and the conqueror's land and riches to man, Indra is a deified myth based on the historical facts of the descent of the Āryans to the sub-continent. Agni in contrast brings the Gods as the "willing ones" to man and he brings down himself (*ṚV.* I; 44, 8), Agni is the one who burns the bad things (*ṚV.* I; 12, 5), he is the sage, the purifier, just and good. Agni is described as "our messenger (to theGods)" (*ṚV.* I;12) and he is source of inspiration, spiritual fire (*ṚV.* I; 21,3). the *hotṛ*, the wise, master of wealth: intellectual, spiritual and material. Agni is bearer of oblations, much beloved, whose ordinances are true, resplendent, chastiser of mischievous people, sage lord of the house, guardian of virtuous people, ever young and ancient at the same time, purifier and with life-splendour.

Agni is known from time immemorial (*ṚV.* VIII; 75, 1) and is invoked to bring forth valiant men and food, material, intellectual and spiritual, as "food for thought"; he is the giver of joy and gives room, for instance by clearing the forests by fire (*ṚV.* VIII; 75, 11). Agni is seen as manifested in nature's glories and as a source of nature's gifts, while Indra is the distributor of man-conquered wealth. Agni, the enlightened and giver of enlightenment is called *Jātavedas*, the all knower and the great *Asura* rich with magic power (*ṚV.* V; 12, 1—*ṚV.* V; 15, 1). The dual divinity Indra-Agni associates Agni and

Indra because, in a way, they complement each other.

Agni has a few distinctive appellations that are exclusively his; such are *Tanūnapāt* or son of himself (*ṚV.* I; 13, 2- *ṚV.* I; 142, 2 *et al.*) because fire is lit by fire and the concept of God implies that God **is** but cannot be generated by any other entity. Another significant appellation is *Narāśaṁsa*, the praise of man (*ṚV.* I; 13, 3 *et al.*), or the one who has shows the way (*ṚV.* I; 18, 9); Agni Vaiśvānara is beloved of all men because he gives supreme knowledge as well as practical usefulness that provides better good, better implements, and better weapons. Agni is the one who prepares the sacrifical food which in turn can also be interpreted, I believe, as "the understanding necessary to attain knowledge", because knowledge is the food of the mind; "where on the *barhis*, the sacrificial grass, immortality is beheld" (or: can be reached, *ṚV.* I; 13, 5) and where the gates of Ṛta are opened to the mind (*ṚV.* I; 13, 6). Witnesses to this are *Ilā* (nourishment), Saraswatī (flowing waters) and *Mahī* (the great one, *ṚV.* I; 13, 9) who are givers of comfort for body and soul. Tvaṣṭṛ, the skillful architect and Agni are invoked to be present (*ṚV.* I; 13, 10) for the understanding and use of nature. Throughout, Agni is the directing intelligent mind; as the *hotṛ* he ensures communion with the Gods, he is always the lord of the house and always kindled anew with lively flames.

The idea of God in early man would have been a mixture of wonder, fear and interest to obtain help, support and practical benefits. Agni is both intangible as God and utilitarian, as fire. Agni as everything else has multiple manifestations and plays many roles; he is character and mind purifier (*ṚV.* I; 13, 1) and also, as fire, the means to clean infected wounds, purify the house and stable and burn away all pollution and disease; fire is used for smelting and purifying metals from their ores, to make steel and extract active plant substances by crushing, boiling and distillation (*ṚV.* I; 36). As Purifier par excellence, Agni is present in the house fire and at all *Vedic* sacrifices, he brings physical and spiritual food *(ṚV.* I; 31, 11). Agni God and common fire are distinct (*ṚV.* I; 26, 7): "may he be our dear household Lord. . . we with bright fires be dear to him" (Griffith transl.) or "may the Lord of men . . . be dear to us and may we also be dear to our Lord by our spiritual fervour" (SS and SV transl.); in *ṚV.* I; 36 "the ancient" may be Lord Agni and the "young one" the newly kindled fire, But either as God or as common fire, Agni is everywhere and always present, unborn as all-pervading potential energy, or born by friction (*ṚV.* I; 67) or otherwise manifested. Repeatedly the creative power of Agni Vaiśvānara is invoked (*ṚV.* VI; 9,3 *et al.*).

Fire has been a boon to man (*ṚV.* II; 1, 2) who recognizes (*ṚV.* I; 26, 8-10) that the "Gods adored with brilliant fires have granted precious wealth to us. So with bright fires we pray to thee" (Agni). . . "with all thy fires find pleasures in this our sacrifices and this our speech, O son of strength" (Griffith transl.). "Blessed by you men with spiritual fervour imbibe the best of your qualities; may they use them to help mankind" (*l.c.*). "O . . . venerable God, source of strength, may you with your life giving energy . . . guide us with sacred wisdom (*ṚV.* I; 31). This vividly conjures the image of the beginning and evolution of the thought of Agni as God, the one who is frequently mentioned as the original God "the best and earliest Aṅgiras". "a God among Gods" and where the mythology in support of the statement follows: "Agni, revealer of the *Śruti* to mortal men and giver of wisdom".

The magical aspect of Lord Agni may have been represented by the mytical being unclearly described as a "divine horse" (meaning, I believe, **power**) called Dadhikrās or Dadhikrāvan to whom fire hymns are dedicated (*ṚV.* III; 20- *RV.* IV; 38. 40-*ṚV.* VII; 44) and "from whom came the gift in early times". Dadhikrās has been identified b ycommentators with the morning sun, which cannot be correct since he is "compared" and "likened" to the sun, not **identified** with. In fact Lord Agni as God is never fully understood (*ṚV.* V; 19. 1).

Fire is red and yellowish and to this day red is an auspicious and well liked colour and ochre-yellow or red is the colour of the robe of *sādhus* and *lāmās*; at Holi, red is the most widely used colour and, of course, it is the colour of a bride's attire. Agni Vaiśhvānara is the friend of man (*ṚV.* I; 59) in that he is not aloof (*l.c.*) and he is always at hand to help man as lord of all bounties, plants, animals and minerals (?) literally what is mentioned is: "riches in the mountains" which could be mineral ores. The Arāvalli hills were mined for copper ores as far back as ^{14}C-dated more than 3000 years BP, and "sustain man as deep planted pillar". As common fire, Agni exists (*ṚV.* IV; 33-37) to satisfy the Gods with hard work with the approval of the great ancient architect Tvaṣṭṛ. The legend regarding the deft-handed Ṛbhus are very old and the Ṛhbus themselves are deified humans associated with the Ādityas and experts in-"the ancient sacrifices"; the hymns related to the Ṛbhus and to Tvaṣṭṛ by Sage Vāmadeva are indeed pure science and technology in poetic form.

Maybe a word would be in order here to further distinguish between Agni as fire, that indicates strength, Agni as Sūrya that indicates energy, and Agni as Śakti, that indicates power, as can be deduced from the context of the hymns of the *ṚV* that taken together constitute powerful internal evidence.

In conclusion Fire appears as heat, light and flames and as "other fires" (*ṚV.* I; 59, 1) including other forms of energy such as life processes and basic technologies. The living body is warm, the sick body is hot and the dead body is cold. Cooked food is better for health and makes stronger people, the tip of sticks can be hardened by fire and fired pottery is harder and lasts longer than sun-baked earthern pottery or bricks. Accordingly I tried to find out whether there is in the *Veda*;

1. discrimination between fire and Agni;
2. evidence of the use of fire in science and technology; and
3. mention of the antiquity of Agni as God.

1. Discrimination between common fire and Lord Agni: The omnipresence of Agni is repeatedly stated and the omniscience of Agni Jātavedas is a recurrent theme; consequently, as discussed above, Agni **Is** God, the primordial God (*AV*, III; 20, 21) differentiated from common fire that was made Agni Vaiśvānara. On the other hand, common fire is a manifestation or an expression of Lord Agni and this is also how the Zoroastrians see fire. As fire, Lord Agni has many ways of expressing himself, but fire for technical purposes is respected and eulogized through the Ṛbhus.

According to the Schuré (1899) the great scholar of esoteric traditions, Agni is the first born among the Gods, the lord and generator of sacrifices, the conveyor of all mystical speculations, he has sacrificed himself as his object, generates other Gods, organizes the world, produces and maintains universal life and is a cosmogonic power. It seems to me that though not in the correct chronological order, Schuré describes the "career" of Lord Agni, from his first birth as God in the mind of early man to the Jātavedas of the *Veda*, and the energy/matter concept of modern physics.

Common fire is a natural phenomenon and is used in many ways as "he who travels in various directions"; in fact man mastered the art of making and using fire in many ways since hundred of centuries ago. Physically Agni burns, spiritually he destroys evil thoughts and intellectually he provides for acquiring and keeping knowledge (*ṚV*. I; 79). But Lord Agni as spiritual fire (*ṚV*. I; 19, 2) is beyond complete comprehension and perfect mental representation. The name of the God and of common fire may have remained the same, but the concepts of Lord Agni encompasses much more than just common fire. He is terrific (*ṚV*, I; 19, 5), both benevolent and protective and agitator of major forces such as clouds, rains, oceans (*ṚV*. I; 19. 8) and like Savitṛ, among other means he expresses himself through Sūrya.

Finally the dual aspect of Agni God and Agni fire is clear in *ṚV*. I; 99, where *Jātavedas* the all knower is invoked to consume as fire "the wealth of the enemy and as God to carry us through grief and troubles as a boat across the river!". Fire destroys the forest, but also clears the way for agriculture and pasture land; he protects as a guard and is inspirer of sacred actions and promotes prayers. Agni lies in the innermost core of the heart (*ṚV*. I; 145, 5) where the soul resides, but man depends also on common fire for practical purposes of everyday life (*ṚV*. I; 150).

2. Evidence of the use of fire in science and technology: Internal evidence in the *Veda*, shows that the knowledge of the origin of the use of fire by *Vedic* man and by the other men with whom he came in contact, was lost in the ancient past. I mentioned above the triple meaning and triple nature of the *Vedic* language. Two such meanings are usually easy to perceive, one is material and the other spiritual. The third or divine and esoteric meaning is meant exclusively for the education of the initiated few, it is spelt out as mnemonic series of catchwords rather than explanation of a subject that would already have been discussed orally at length during the period of teaching and indoctrination of the *śiṣyas* by the *guru*. The esoteric meaning is subject to different interpretations and may have been formulated differently by various commentators, philosophers, scholares and translators. It must be noted that even where purely technical uses of fire are intended, Lord Agni is invoked before any major, or minor, enterprise is undertaken. This practice is quite natural since there is no distinction between sacred and profane. The fundamental concept of the Indian tradition that everything

is made sacred by the omnipresence of Agni God is ecologically valid; fire and God Agni and their respective ecological roles are perhaps nowhere else described as well as in *ṚV.* I; 65 to 73, specially hymn 67.

3. The antiquity of Agni: Very early (*ṚV.* I; 31-*ṚV.* I; 189) fire worship was instituted and already in the *ṚV* it is recognized as an ancient practice (*ṚV.* I; 36, 4 - *ṚV.* I; 96, 1-3 - *ṚV.* I; 127, 10). Agni is said to be the "first to come among the divines"; Agni is "the early Aṅgiras, a God and God's auspicious friend" and " the sages were born after thy laws, active in their wisdom" (M. Müller transl.). Earlier commentators (MM and Oldenberg, p. 53) saw (*ṚV.* I; 60, 1) reference to the celestial **and** terrestrial birth of Agni, who was brought to Bhṛgu and Mātariśvan as well as reference to the mythical ancestors, the "*ubhayāsah*" who first established the sacrifice to Agni.

In *ṚV.* I; the expression of Agni through Sūrya has already matured fully in the Agastya hymns and at the end of the *Saṁhitā* (*ṚV.* X; 183, 3) Lord Agni is established as the giver of life and energy and **He** expresses himself "in plants and herbs, in all existing being I have deposited the germ of increase. All progeny of Earth have I engendered, and sons in women who will be thereafter" (transl. Griffith).

I would refer back to Dandekar (1979a) who draws attention to the spiritual aspect of the Vasiṣṭha hymns and who sees the origin of *bhakti* expressed in the powerful addresses of the *ṛṣi* to Varuṇa. Dandekar also shows that the *asu* or significance of Varuṇa can be understood as the "one who possesses magical power". It is possible to perceive a strong belief in absolute monism throughout the VII *Maṇḍala* of the *ṚV.* , where Agni personifies knowledge - abstract and practical and Varuṇa personifies the unintelligible and unexplained aspects of the universe; for this reason Varuṇa operates through and is magic itself. The *ṛṣi* firmly believes that "no secret can be withheld from your knowledge"; (*ṚV.* VII; ; 61, 5) in a hymn addressed to

Mitra-Varuṇa, the practical/theoretical pair, the direct and indirect knowledge, the day-to-day reality of the world in which the *ṛṣi* lived; but the *ṛṣi* also believes in the ultrasensory reality that he sensed as the power of Varuṇa, which is above man's comprehension.

It is possible to render most, or much of the VII *Maṇḍala* in plain bio-ecological terms, since "men in deep thought engendered glorious Agni (*ṚV.* VII; 1, 1) . . . Lord of the homestead . . . to be honoured in the home forever (*ṚV.* VII; 1, 2) friend and granter of wealth and wisdom (*l.c.*; 1, 5), far seeing (*ṚV.* VII; 4), wise and generous, **who has allowed man to seize him**. The words indicate the gratefulness of the thinking man, the scientist who becomes one with God, which **is** that which he has understood. Further the "*ṛṣi* is to bring help to men" (*ṚV.* VII; 26, 5) from what he has understood. However, at this stage the identification of man with the Godhead is qualified, because only the **consecrated fire** is Agni God.

The syncretism of the Vasiṣṭha hymns lead to a naturalistic monism that pervades Indian philosophy and the Indian way of thinking and living that culminated with Ādi Śaṅkara. Much later in Italy it found expreession through Giordano Bruno who expressed comparable ideas of a general, all pervading, naturalistic pantheism. An example of the inter-relationship that the *ṛṣis* saw in all aspects of nature is expressed in *ṚV.* VII; 12, 13 where "Agni, thou art Varuṇa and Mitra; Agni is the great God that comes from all directions" and has all meanings and values; Agni is eulogized as the enlightner of all, who has filled heaven and earth with his energy and knowledge. Agni both taught how to pray and gave practical advice.

The first hymn of the third *Maṇḍala* (*ṚV.* III; 1 by *ṛṣi* Viśvāmitra) can be read as the evolution of Agni from fire and sun into food; it also indicates growing complexity of rituals that develop to satisfy the striving for perfection and the desire of man to always surpass himself. Perfectionism in this case aims at more and more powerful worship of Lord Agni and the acquisition of magical powers. It is the same *furore* that struck Michelangelo Buonarroti who left his masterpieces unfinished and his body exhausted because he could never attain the level of perfection that his own super-ego demanded, nor could he attain the beauty of the God in which he devoutedly believed. The achievements of giants like *ṛṣi* Viśvāmitra, *ṛṣi* Bharadvāja, *ṛṣi* Vasiṣṭha or Michelangelo were not by them considered to be achievements, they considered them only as a partial unveiling of Ṛta; they were always unsatisfied with their own work and were tormented by a feeling of falling short of their high ideals, of being unworthy and unable to muster the perfection that they demanded of themselves but that they saw only in God.

It is not only in India or in the Āryan or Indo-European world that fire is an omnipresent reality. To this day man-made fire is kept to honour and to be honoured in homes and places of worship, by people of many cultures and is often taken as a manifestion of God.

However, as an afterthought, a sad remark: when man learned how to make fire, he likened himself to God as he conceived it; then as the appprentice sorcerer that man is by nature, he became afraid of his own creation and had to propitiate it "at *Iḷā's* holy place". The beginning of man's pride is evident in *ṚV.* I; 95, 9: "O fire divine, kindled by us . . . " and in *ṚV.* I; 127, 8 where fire is called the "guest of man". Rituals therefore had to be developed also to appease man's own fear and to request boons. With time, inevitably, rituals became more and more complex: "may we improve the rites with new performance" (*ṚV.* I; 31, 8) and finally, they may become an end in themselves: "this very new worship as it is performed (*AV.* XIX; 43, 1); "O Agni thou wast born for our purpose" (*AV.* I; 7, 6).

In another context, Dandekar (1979a, p. 7) proves that rituals came later and younger members of the *Vedic* pantheon are more prominent in *Vedic* rituals than older Gods as Agni or Savitṛ themselves.

Undoubtedly *Vedic* Man perceived that all light on Earth is the outcome of combustion, which is fire or Agni in his many different manifestations. Agni expresses himself also as enlightenment of the spirit and of the mind. The light of perception and understanding are a result of Agni as Energy. The cold light produced by phosphorescent animals such as fire-flies or fire-beetles, is the exception rather than the rule. Man the Inventor was able to invent cold light, also called tube light, only as late as the XX century and that too, as the result of a practical joke.

In conclusion, all light, far or near, directly or indirectly, comes from Lord Agni. *Vedic* man saw no other light.

"By Thee, O Agni, Varuṇa who guards the law, Mitra and Āryaman, the Bounteous are made strong;
For as the felly holds the spokes, thou with thy might pervading hast been born encompassing them round".

ṚV. I; 141, 9 (transl. Griffith)

"Agni was born the first of all, like Sūrya lovely with his light; with lustre these have spread abroad."

ṚV. VIII; 7, 36 (transl. Griffith)

6

Load Savitṛ, Life and Hope

"Let no one here debar me from enjoying the
golden light which Savitṛ diffuses
He covers both all-fostering worlds with praise
even as a woman cherishes her children".

ṚV. III; 38, 8

IN *Ṛgvedic* times there probably was no differentiation between the spiritual and the intellectual, as there was probably no differentiation between knowledge and Truth. Truth was knowledge and Intellectual was Spiritual. This is one reason why ideas could be as concrete as what we now call "real" and this is also one reason why I believe that abstract ideas, or ideas/concepts and deified ideas would have preceded the personalized forms of anthropomorphic divinities. It is what in brief I stated as "philosophy precedes religion". Thus, coming back to my initial proposition that the ancient-most God or Gods are personifications of ideas/concepts, I see Savitṛ as the very ancient characterization of the abstract concept of Life, "the life in living", the **it** that makes **living beings** a category apart. As "fireness in fire", the idea of "life in living" once it was conceived, must have been one of the most bewildering thoughts man ever had.

Savitṛ the God is often mentioned in association with Bhaga, a deified man, the Āditya who brings good fortune and therefore gives Hope.

It appears logical to speculate that early man must have wondered frequently over the magical miracle that is life, and the generation of new life. First as an idea of something that exists but is intangible, like the energy and the power of fire, then as a concept of something that makes a body different before and after death, the wonder of **living** and the concept of **life** must have developed into the characterization of Lord Savitṛ. The concept of Hope is implicit in the concept of Life; "for my protection I invoke the golden handed Savitṛ" . . . "praise the water offering, O Savitṛ" . . . "we call on him . . . on Savitṛ who looks on men" (*RV.* I: 22, 5-7). The concept of Life is different, unique and means more than the concept of energy; thus Agni is the energy in nature, he is powerful and strong in his physical aspects (*RV.* I; 94, 12), he can sooth the wrath of Mitra, Varuṇa and the Maruts, because they are all dependent on him, and he intercedes for the natural forces of water, earth and air to protect man; in other words the powerful Agni is to be invoked for protection against the unleashed fury of nature against which puny man is helpless, but Savitṛ who is **always gentle** and not particularly strong is otherwise extremely powerful since he is that something, the indescribable life-principle that uses the energy and power of Agni to generate and maintain life in living beings. Early man recognized that the life-principle is the same in all living beings. Without the active presence of the "life principle", Agni himself would be unable to create and sustain life. It is abundantly clear that "energy" is one that moves the cosmos in its different aspects, including the world we live in; this energy is equated with Lord Agni, the essence of fire; only Agni can protect man against himself, but only Savitṛ can impart, create and sustain life and only he can put in motion what can move, including astral bodies who are imbued with a life of their own, the explanation of which is inaccessible to earth-bound man.

Savitṛ as the essence of being alive, controls and directs living beings; in other words that undefinable, non-material, unbound principle that makes a living being different from a dead body, is Savitṛ. Savitṛ is compared to a mother, the original human mother Aditi, who is the unbound, unfettered who vivifies, stimulates, regulates movement and life processes. Savitṛ similarily impels life processes and the act of living in the right direction and of course he is in close association with Ṛta that is all the law and order of the cosmos, inclusive of the law of all matter, ways and means that make life possible. Savitṛ in the *Gāyatrī mantra* is associated to, but is not identical with Sūrya who is at the root of the web of life on Earth. "Of Savitṛ the God, the sapient *Asura*, we crave this great gift which is worthy of our choice" (*ṚV.* IV; 53, 1 - transl. Griffith).

Savitṛ is the personalization of the concept of the most impressive magic, the one that makes the act of living possible; he is the concretization of the very difficult to understand and impossible to explain idea of **Life**, that, unlike fire, man is unable to create. **Life** is a riddle that man has been unable to solve to this day in any of its multiple aspects. Life may well have been seen as an act of magic of *Asura* Varuṇa and Savitṛ himself would be imbued with Varuṇa's *asu*. Life is the most precious possession of man, hence the golden hue, the golden hands and arms, the golden tongue, the **multiform** golden and bright chariot of Savitṛ and his association with the golden Sūrya. The chariot that carries Savitṛ, has an infinite variety of forms and is bright as life itself is. Sūrya is then an expression of Agni, the mediator for the realization of life, that begins on earth with the photosynthetic process, active only in the presence of light.

Savitṛ endowed with great *asurya* raises his golden arms, he is motion and life in motion and he distributes these powers among all beings. The "gesticulation of the God is . . . typical of the great world magician producing his magnificient magic" (Dandekar 1979a, p.25) which, as I see it, is **Life**. By the movements of arms and fingers

Savitṛ infuses life, at dawn night is sent away to await his arrival (*ṚV.* I; 113, 1) and he, Savitṛ, awakes all diurnal animals and man while he puts to sleep the nocturnal animals, the process is reversed at dusk when Savitṛ puts to sleep diurnal animals, men and green plants and awakens nocturnal animals, thus completing the most basic cycle, the day-night cycle. Savitṛ, the God true-minded (*ṚV.* I; 73, 2), also puts to rest, temporarily and permanently when he spells the end of living and the advent of death: "for thou art might to produce and lull to rest the world of life . . . " (*ṚV.* VI, 71, 2).

Savitṛ has a well marked personality within the Varuṇa philosophical complex, as discussed by Dandekar (*l.c.*, 1-27) and guided by my stated biological bias I concluded that the beautiful and mysterious figure of God Savitṛ is the personalization of the idea of Life and Hope. God Savitṛ is the Enlivener, Vivifier, Supporter, Impeller, Luminous, throughout the *Veda* the aspects of Savitṛ as motion are emphasized, and life is motion. I see Varuṇa encompassing all **magic,** all the *asu*, all that in the universe is not understood or explained by man and I see Savitṛ as the abstract idea of Life; Savitṛ represents that aspect of Varuna's *māyā* that refers to the presence of all life on Earth and that he, Savitṛ as the Grand Stimulator and Vivifier, "puts in motion". Savitṛ the Lord is Life itself, and Hope, represented by Bhaga, his regular associate: may Bhaga and Savitṛ send us forth, for instance *ṚV.* V; 42, 5-*ṚV.* V; 82, 1- *ṚV.* VII; 38.1.

As mentioned above (Chapter 3) magical powers are invoked and become necessary to represent mentally those phenomena that are not otherwise obvious to the intellect and as Dandekar points out, Savitṛ transcends the limitations implied by the nature of Sūrya and, I would add, of all other Gods with whom he has been identified at one time or other by different commentators.

The perception of the existence of an imponderable, invisible, barely perceptible **something** that now we call "life" and the formulation of the idea of life as a moving principle or acting idea, the formulation of the working concept of life is an intellectual feat as great as the formulation of the concept of *ākāśa* or, later on, of the concept of *śūnya* or zero.

Savitṛ is as much the life principle in men, women and children as he is the life principle in all animals and plants as well as in the unseen, but well manifested in their effects, living things that multiply and turn milk into curds or cause the fermentation of plant juices; they all grow, multiply and generate something similar to themselves.

According to Dandekar, the word Savitṛ derives from the root *sū* (to stimulate, to vivify) and "the activity suggested by the root *su* is very often characterized by some form of mystical or magical power" . . . "*sū* denotes . . . some form of magical underlying a sort of magical charging" (Dandekar, *l.c.*, p. 23). Savitṛ therefore is not only the Stimulator, the Vivifier, the Initiator, he is also somewhat mysterious as life also is. The *Vedic* approach to and representation of the mysteries of life is logical; the mysteries of living and dying have been a constant preoccupation of men of all times. Undoubtedly life is the greatest of all magics and mysteries and the one of most direct concern ever.

Dandekar (*l.c.*) published a thorough study of Savitṛ and convincingly showed that the "essential nature" and personality of Savitṛ "is only an aspect of Varuṇa", that "the movement of Savitṛ's hands and fingers is particularly emphasised by the Vedic poets" . . . and that Savitṛ of the *ṚV* is a representation of Varuṇa in his special gesture as commandant and the magician of the world, a gesture which is easily the most impressive and at the same time the most expressive of the essential personality and nature of that God". However, it appears to me that Savitṛ is more than just a representation of Varuṇa, he has a reality of its own as the grand Vivifier and since only Savitṛ can "put in motion" or give life, he must be the essence of life itself.

I should like to point out that essentially my reading does not differ from the interpretation given by Dandekar, all the references that the master quotes agree also with my reading; the latter, however, adds one more dimension to the description of the personality of the Great Savitṛ. My interpretation expands the God's attributes and characteristics, it expands its meaning and concept as the life principle that pervades all those components of the universe whom we recognize as "living". My reading upgrades, if I may say so, Lord Savitṛ to be the **essence of Life** itself. In the reading of this biologist, Savitṛ is first and foremost the Vivifier-the one who gives life, because it is he who **is The life-principle** itself; Savitṛ is therefore the source of Life, the same life that he ceaselessly regulates. Each line, each verse of the Veda bear literal, intellectual, spiritual, mythological, symbolic and figurative meanings, all of them valid, specially when dealing with such complex a phenomenon as Life; thus, none of the possible readings necessarily invalidates earlier readings that were "decoded" with different "keys". I have used a bio-ecological key.

Dandekar (*l.c.*) also showed that Savitṛ is a very ancient God, he is associated with the older Gods and can be traced back to the proto-Āryan ideology; Savitṛ is not, as some authors supposed, a newcomer to the *Vedic* pantheon. This conclusion confirms my opinion that the original God or Gods are born of abstract ideas that evolve into concepts that later become anthropomorphic and are finally encompassed in the religious canon by rites and rituals. "The Wise One arrays himself in every form: for quadruped and biped he hath brought forth food" (*ṚV.* V; 81, 2).

All Gods are alive, hence they cannot resist Savitṛ's will. Savitṛ, genderless like Agni, perhaps combines the attributes of both sexes because the act of living, the essence of life and the physico-chemical processes that sustain life are sexless and irrelevant of sex-attributes;

living is a magic in itself, exclusive and never duplicated, accordingly I see that Savitṛ "includes and transcends the distinctive features of all these Gods" (Dandekar, *l.c.*, p. 15). Sūrya visibly and physicochemically puts life to work through photosynthesis, while Savitṛ is the subtle, imponderable principle that enables Sūrya and Agni to promote life processes that can be gauged, for instance growth, maturity, reproduction, senescence. Sūrya is seen correctly as light, Agni as heat or energy in general (*ṚV.* X; 114, 1) while Savitṛ remains indefinable; "O much praised vital principle, O Savitṛ" . . . "when the adorable Lord . . . mounts high . . . then bipeds and quadrupeds and the flying birds in the fulfillment of their ambitions are set to their tasks" (*ṚV.* VIII; 27, 12) (SS and SV transl.) or "May this God Savitṛ, the Lord, the offspring of waters, pouring down his dew be gracious" (*ṚV.* VI; 50, 13).

God Savitṛ as the Enlivener is asociated to water and air, or wind, *prāṇa*, without which the life-principle cannot express itself nor become a concrete reality; life cannot be put in motion or exist without water and air (*ṚV.* II; 38, 2). Water is offered to Sūrya, when greeting him at dawn and Savitṛ drives away immaterial evil spirits and he is sinless.

In a comparable manner we may conclude that Lord Savitṛ can give and withdraw life of living beings; he is the **it** that makes a living body different from a dead one, the **it** by which Lord Agni can act, the **it** that generates, sustains and destroys life; (*ṚV.* I; 150, 1). "Most gracious God who brings to life and lulls to rest, he who controls the world, what moves not and what moves. May he vouchsafe us shelter-Savitṛ the God-for tranquil life, with triple bar against distress" (*ṚV.* IV; 53, 6). "Full of wisdom . . . the God stretched out golden arms that he may bring forth life" (*ṚV.* VI; 71, 1).

Savitṛ is often described in terms of life cycles (*ṚV.* II; 38, 4 - *ṚV.* III; 33, 6) such as circadian rythms or the awakening after night's rest; he conrtrols and separates time intervals and the duration of cycles and parts thereof, like the season and day and night: " . . . the night obeys Savitṛ" and he is called when the soul is ready to depart.

As life-order, Savitṛ is praised and, significantly in the same hymn (*ṚV.* I; 22, 9-12) Agni is requested to bring the spouses of the Gods and heroes, who are individually named and who, evidently, indicate the feminine role in the perpetuation of human life. In other words, thanks to the energy of Agni and to Savitṛ's magic, life can be maintained and regenerated through feminine power and intervention. But, since there can be no life without water "rich in wealth", water is also invoked (*ṚV.* I; 22, 14) in the same context. Waters or liquids hold all medicines and hold, potentially, the energy of fire (ṚV. I; 23). Savitṛ's association with Agni is clear (*ṚV.* I; 95, 7) and the latter is said to "stretch his arms (flames) with might **like Savitṛ.**"

Savitṛ the God true-minded (*ṚV.* I; 73, 2) has also an ethical aspect since he is said to make man sinless, which means disease free, and therefore man should not become a hazard or handicap to his family and the society at large; the expression of making man sinless can also be read as an encouragement given to man to use life properly. Illness and poor health were regarded as punishment for transgressing Ṛta or as the result of a curse that might have been well deserved. The sun is Savitṛ's *Sūryaśmi* as he is also mentioned as Varuṇa's eye. Savitṛ bestows immortality to the Ṛbhus for their good deeds (*ṚV.* I; 110 , 3) and to meritorious men (*ṚV.* IV; 54, 2, 5-6) while Varuṇa guards the immortal world of the non living. Savitṛ binds living beings with the laws that control and regulate all the cosmos from the movement of the celestial bodies, believed to be alive because they move, like Sūrya himself and which are all included in Ṛta, to the invisible creatures that make curds and other foods.

Finally I would recall that it is **Savitri,** daughter of king Aśwapati, a name that is probably a symbol of all Āryans, who outwits Yamarāja himself and causes life to return to the dead body of her husband Satyavān, the truthful.

The reading of the hymns dedicated to Savitṛ brings to the mind's eyes the image of puzzled *ṛṣis* thinking about "what is life?"- "what keeps it going?" - "by what magic does it develop and evolve?" - "is being alive the same in what moves and moves not?" - "what is this it whose absence spells death?"

"Savitṛ God who looks on every creature
laying to rest the immortal and the mortal
. . . he, Savitṛ gold handed *Asura*, kind leader"

RV. I; 35, 2 and 35, 10

"For thou (Savitṛ) at first producest for the holy
Gods the noblest of all portions, immortality;
Thereafter as a gift to men,
O Savitṛ, thou openest existence, life succeeding life".

ṚV. IV; 54, 2 (trasl. Griffith)

7

Conclusion

Contemporary and *Vedic* Expression of Ecology

THE word "Ecology" is a relatively recent addition to the scientific vocabulary, it is not much older than about 100 years. It literally means: the science of the house. More specifically, Ecology is the description and analysis of the structure and dynamics of the environment in which living beings exist and of which they are a part. The inter-relationships of living beings among themselves and with the environment as well as the laws that govern these processes and interactions are the subject matter of ecology. The environment in which biological species live includes both living and non-living things, therefore all the factors that influence the existence of life and reproduction of plants, animals and microbes are the object of ecological studies.

Andrewartha, *et al.* (1954) defined ecology as the study of the distribution and abundance of individuals of biological species. Obviously the presence or absence and the number of individuals of any species at a given place and time is the result of the sum of the positive and negative factors that favour or prevent the life and reproduction of that particular species at that particular place and time.

How did ancient man perceive the relationships of cause and effect of living beings among themselves and with their environment? Present day ecology goes into details of physical, chemical, biological, socio-economic, climatic, geographic and other aspects of these relationship and interactions; how far did *Ṛgvedic* man progress towards identifiying these relationships? Did man in those far away days develop a methodology adequate to analyse processes and to apply the results of the analyses to practical purposes? How far did man go in altering the functioning or dynamics of natural ecosystems? How was the socio-economic structure of the community of man influenced by climates and ecological features and how did he react to them?

From the previous chapters it is clear that *Ṛgvedic* man did not formulate ecological or any other scientific theory in the contemporary sense of the word "scientific theory". The question is, did he perceive the principles on which scientific theories were later based? How much was his culture influenced by the different environments that he occupied?

The answer to these questions is that he did study the environment, that he drew practical conclusions from his studies and that his culture was, as all other cultures, strongly influenced by his environment, indeed he reacted to it by developing a way of life systematized down to the minutest daily details of the activities and habits of the individual, of the community, of man's participation in the cosmic drama and in his religious beliefs and life. The fundamental questions that ancient man put to himself would have been related to what we now call matter and energy, matter in general and energy as power; to man's place in nature; to life, death and disease; to immortality and the means to satisfy his own needs. With time also questions such as "where am I?", "who am I?", where do I come from?" would lead to philosophical and religious specultations. However, the first observations and lessons that man learned from nature would have been related to what we now know to be the fundamental questions of the present day science of ecology, they were matters of survival importance, related to matter and energy and to space and time.

The single most important ecological principle is that of "competitive exclusion". According to this principle two species cannot occupy simultaneously the same ecological niche indefinitely; this is where enters the concept of space and time. This principle was first formulated mathematically by Volterra (1926, 1928) and later elaborated biologically by Gause (1934, 1935), Hutchinson (1959, 1961), Margalef (1956, 1958a, 1958b), Odum (1959) and other ecologists, including myself. Ancient man could not be aware to the subtle reason why two species cannot coexist indefinitely in a limited space and time under constant environmental conditions, though he could easily see that two material bodies cannot occupy the same place in space at the same time. Matter and energy are more rapidly understood than space and time. Man perceived life as a constant struggle and as such it is often mentioned in the *Veda*, however it would have been difficult to perceive that each individual and each species lives in an hypervolume where the vectors representing ecological factors unremittingly change in magnitude and direction. The ecological niche concept is better understood as an idealized hypervolume of conditions that endlessly changes form and dimensions. Another reason why this would be difficult to perceive is that unending changes in the magnitude and duration of ecological factors cause both a short term and a long term variability of the environment of each and all ecological niches at all times. Uninterrupted and continuous changes of every ecological factor along time is perceived as movement and growth, phenomena of which ancient man was well aware of as he was also aware of the continuous change and evolution of everything. In this sense the *ṛṣis* and all ancient thinkers would be, in contemporary ecological terms, ecologists of the

Andrewarthan school since they would see the changes of the distribution and abundance of individuals of the different species in space and time, even without knowing or understanding how this would happen and what would be the underlying causes. Ancient man would be aware only of the end result of the dynamics of ecological factors that he would otherwise be totally ignorant of.

The main conclusion of the theory of competitive exclusion which is mathematically correct, would seem to be empirically false, unless we realize that constant actions and reactions take place between an infinitely large number of ecological factors. The whole is extremely complex and unless we undertake strict scientific analyses, we would perceive only the end results but no details. The theory may be proven right but its empirical aspects were the only ones that early ancient man could in fact see and sense from his own participation in the dynamics of the world around him.

In fact *Vedic ṛṣis* saw harmonious movement everywhere, they saw constant changes within the harmony of nature and this they interpreted as the law and order of the universe, Ṛta. It is in this sense that celestial bodies, rivers, forests are living beings, move around and grow and reproduce and thus participate in the cosmic drama. Matter and energy are everything down to the atomic levels; it is the interaction of the two members of this couple that originates life and maintains it. This active and effective principle man saw in Fire, Lord Agni. He recognized it in the five *mahābhūta* or elements of existence: earth (solid matter), water (liquid matter), air (subtle matter), fire (energy) and *ākāśa* that is the other side of the coin (void, non-existence, the opposite of matter).

The ancient *ṛṣis* put to good use the knowledge of nature gained through empiricism and experimentation as well as that borrowed from other cultures. By and large the use that the *ṛṣis* made of nature, of good Mother Earth and of cleansing water, was wise. Large scale degradation of the environment by over-exploitation was avoided thanks to the small number of human beings that was then well below the carrying capacity of the ecosystem; also some forms of management of the environment were developed through irrigation, forestry practices and animal husbandry as well as agriculture. The deep respect for the life principle personified in Lord Savitṛ and inherent in all living beings, made man one of the many knots of a very complex web, rather than the centre of a universe created for the use and benefit of man himself, as vouchsafed by later creeds. Man saw himself as one of the many elements of the life web; though to his own eyes he was the most important of all he did realize that he had to attune himself, his behaviour and his actions to the eternal law of the Universe, this is *Dharma*.

The attitude of identification of ancient man with the world around him as taught by the ancient *ṛṣis* is evident in the *Veda*, it reflects man's perception of the structure and dynamics of nature or what we would now call the ecological aspects of the many different environments colonized by man.

Glossary

Ācārya	teacher, guide, preceptor of spiritual knowledge and martial arts; Droṇa was the famous *ācārya* of the Paṇḍavas.
Adhibhautika	literal, material or social interpretation of the texts.
Adhidaivika	godly, universal or cosmic interpretation of the texts.
Adhyātmika	inner, spiritual, subjective reading of the texts.
Aditi	infinity, free from bonds, free from Varuṇa's fetters; mother of the Ādityas.
Ādityas	the most ancient gods, Varuṇa, Mitra, Āryaman, Dakṣa, Aṁśa, Dhāṭṛ; sons of Aditi, later six gods were added to the original six.
Agnāyī	the consort of Agni.
Agni	God of Fire and of all sorts of energy, including heat, light, life and reproductin of living beings.
Ākāśa	the first of the five gross elements or *pañca mahābhūtāni; ākāsa, vāyu, tejas, āp, pṛthvi,* according

	to the *Saṁkhya.*
Anārya	non-Āryan people; term also used to mean vile, untrustworthy, despicable, unreliable.
Aṅgiras	one of the seven *mahā-ṛṣi* to whom many hymns are attributed related to Agni and fire rituals.
Aṅgirases	descendants of Aṅgiras; the Bharadvājas are also called Aṅgirases and are considered to be descendants of Aṅgiras.
Apāmnapāt	offspring of waters, an epithet of Agni.
Araṇi, Araṇis	the fire sticks used to produce sparks by attrition, the sparks are then collected on dry grass to ignite the fire; also called the two mothers of Agni.
Artha	all the practical arts and much of what is now known as science and technology.
Ārya	noble, self appellation of the Āryans.
Āryaman	close friend, god of hospitality, together with Mitra and other Ādityas regarded as other aspects of Varuṇa.
Āryan, Āryas	the branch of Indo-Europeans who migrated into India; means faithful, loyal; the opposite is *an-ārya* or non-*āryan.*
Āśrama	the four stages of the life of *brāhmaṇas,* the *brahmacārya,* the *gṛhasthya,* the *vānaprasthya,* and the *saṅnyāsa;* also the hermitage of a *ṛṣi* or sage where

	sannyāsi and *brāhmaṇas* in the fourth stage of life could retire; permanent camps for the education of the younger generation.
Asu	magic power substance or magic potence.
Asura	Lord God imbued with magic power, specially used for Gods who have *māyā*; *asurya*, highest lordship; *asurasya*, asura rule; later use, demon.
Aśvins	twins precursors of dawn, coming on horseback; knowledgeable of medecine and medicinal plants; aśvinau, dual form.
Aswattha	*pīpal*, the sacred fig tree from which the upper *araṇi* is made.
Atharvan	the first man to make fire and to offer sacrifice; a *prajāpati*, associated to Aṅgiras; his descendants, the Atharvana are closely associated to the descendants of Aṅgiras, hence the dual Atharvangirasa.
Atharvaveda	the last and recentmost of the four *Veda*; initially only three *Veda* were mentioned, Manu himself speaks of three *Veda*; partly prose, contains large number of incantations in addition to speculative hymns.
Aurobindo Ghosh	1872-1950, Calcutta; poet, philosopher, mystic; developed what he called a "synthetic philosophy".
Avesta	the sacred book of the Proto-Āryans who migrated from

	Balkh to what is now Iran; sacred book of the Zoroastrians; the *Avestan gāthās* and the *Vedic mantras* have much in common and both have their origin in the Āryan language of the proto-Āryan people; the development of the *Avesta* must have proceeded in a similar way as the *Veda*.
Avidyā	lack of knowledge, ignorance.
Axis mundi	Latin expression, the *axis* that sustains the world and around which it revolves.
Āyurveda	the *Veda* of life, knowledge of life, later considered by some as an appendix of the *Atharvaveda*; medical and medicinal knowledge.
Balkh	the region of Balkh; area where Indo-European tribes settled when they migrated in a S-E direction after leaving their original home in the Ural-Altaï region; they settled in this area for a long time in part of what is now Afghanistan; area inhabited by proto-Āryan people.
Balumain, Balumain-Indra	God of the winter solstice festivities of the Kalash people of North-Pakistan, great intermediate between men and the great God Khodaī, visits the Kalash valleys once a year.
Barhis	sacred grass; several species are considered to be sacred, the most important are the *durba* or *kuśa* whose scientific name is *Poa*

	cynosuroides, *durva* whose scientific name is *Panicum dactylon*.
Bhaga	one of the Ādityas, brings hope and good fortune, closely associated to Lord Savitṛ.
Bhakti	unconditional devotion and surrender due to attachment to God; derived from the same root as *Bhaga*, one of the Ādityas, giver of good fortune and therefore adorable. According to the *Padma Purāṇa* the *bhakti* movement stems from a Dravidian doctrine.
Bharadvāja	an Aṅgiras *mahārṣi* who authored many hymns of the *ṚV*; one of the *saptarṣi*, the seven *ṛṣis*.
Bhṛghu	an ancient *ṛṣi*; in the *Vāyu Purāṇa* his name is added to the *Saptarṣi*, making eight in all.
Bhūta	a being or creature; acquired later, the meaning of ghosts, goblins and often malignant; attendants of Śiva.
Biology	the science of life and of living beings.
Brāhmaṇa	members of the sacerdotal caste; priests are *brāhmaṇas* but not all *brāhmaṇas* are priests.
Brahmā	the Creator of the universe.
Brahman	the Supreme Being, the Ultimate Reality, Self Existent, Absolute.
Brahmacārī	the first of the four *āśrama* in which is divided the life of the *brāhman*; the *brāhman* student, assistant of his preceptor to

	whom he owes undiscussed and unwavering obedience.
Bṛhaspati	also Brahmaṇaspati, the lord of prayer, the heavenly *brāhmaṇa*, the prototype priest, *guru* and counsellor of the Gods; prototype of the *purohita* or family priest and counsellor to kings; also called a *ṛṣi*.
Bṛhat	great, important *Sāman*: the *Bṛhat- Sāman* is one of the most important *Sāma* hymns; from the root *bṛh*, the "stirring up of thoughts" and the fullness of feeling for God.
Capĕre	Latin verb from which is derived the English word capacity or volume of the internal space of a container, cubic content, also mental power, function and character.
Cakra	wheel, circle, said to be innumerable in the human body, they represent centres of energy; the main *cakra*, the *padma*, corresponds to the solar plexus of the human body.
Chalcopyrite	sulphide of copper and iron ore, $CuFeS_2$.
Chaos	total confusion, lack of order; the mathematical word designates a special type of order that can be described as a "deterministic randomness".
Circadian rhythms	oscillations of behaviours and events that follow approximately or precisely a 24 hour pattern;

	events that take place every 24 hours or nearly so.
Conch	vernacular name of the shell of a Gastropod Mollusc whose scientific name is *Xanchus pyraeum*, it applies particularly to Kṛṣṇa's conch, the *pañcajanya*; the oldest archaeological findings of the conch used for jewellry are from the basal level of Mehrgarh, at the foot of the Bholan Pass that links Baluchistan with the Iranian plateau. The fragments of *Xanchus* found there are from the pre-Harappan Period III or Proto-Harappan period, they are dated from the III millennium BC. Many other and later sites have yielded archaeological remains of the conch.
Corpus	Latin noun, body, body of a doctrine, body of evidence etc.
Dadhīca, Dadhikrās	later form: Dadhīca; a *Ṛṣi* son of Atharvan who according to myth sacrificed himself to enable the Gods to made invincible weapons with his bones for the destruction of Vṛtra and the *Asuras*.
Dadhikrāvan	see Dadhīca, the name appears to be interchangeable with Dadhikrās.
Daityas	a race of giant evil demons.
Dante Alighieri	1265-1321, Florence, foremost Italian poet and leading scholastic and political philosopher.
Dards	people of the Hindukush and Himalaya areas, nomads or

	frequently on the move, already known to Ptolemy, Strabo and mentioned in the *Viṣṇu Purāṇa.*
Dāsa	savages, enemies, foes, infidels; non Āryan tribes and people hostile to the incoming Āryans, subjugated by the new-comers.
Dāśarājña, dāśarājña-yuddha	the battle of the ten kings, differently dated about 1900, 2700-1800 or 2700 BC by different authors.
Dayananda	1824-1883; Saivite *Brāhmaṇa* from Kathiawar, philosopher and *Vedic* commentator, joined the *Ārya Samāj* movement.
Delphi	a small temple town seat of the oracle of Phoebus (Apollo) in ancient Greece whose temple was on the shelf of Mount Parnassus, about 600m above sea level, facing South; the Delphic oracle figured prominently in Greek legends, the verdicts were pronounced by the Pythian oracle.
Deterministic-randomness	an uphazard distribution of elements determined by independent non pre-arranged forces. Chaos is a special type of order that appears to be the result of deterministic randomness caused by a multiplicity of impingent forces.
Dharma	duty, justice, virtue, law, righteousness, Duty personified in *ṚV* VIII; 35, 13; the universal law of the cosmos; the law that upholds caste, group and

	individual law, duties and code of conduct.
Dhātṛ	Creator, maker, very ancient God, one of the Ādityas, later preserver of health and identified with Prajāpati or Brahmā.
Diogenes Laertius	4th century BC, the first of the Cynic philosophers who declared to be a citizen of the world.
Dhroṇakalaśa	large container for *soma*, similar ones are still in use to collect grapes' juice for fermentation among the Kalash people.
Dundabhi	drum; war, ceremonial drum.
Ecology	the science of the environment where all living beings find their home, the study of the exchange of matter and energy among living beings and between them and their living and non-living environment.
Eleusis	very ancient temple-city in Attika not far from Athens, seat of the cult of Demeter, an agricultural Goddess; the Eleusis mysteries were imported into Greece and marked the introduction of wheat cultivation; the Eleusis mysteries were held during about 2000 years and were considered to be "excellent" by Cicero himself.
Enzime	organic catalyst, a ferment, substances produced by living cells of plants, animals and microbes capable of varying the rate of specific chemical reactions, usually speeding them;

	they promote the majority of chemical transformations including bodily dissolution after death; the majority are proteins, ex: amylase acts upon and hydrolyzes molecules of starch.
Equinox	spring and autumn day when day and night are of equal length.
Fire	combustion; as God it is Agni, as common fire it is also called agni; Agni is object and means and aim of worship.
Forma mentis	a Latin expression to signify how the mind is shaped, how it reacts and how it behaves, how it operates and how it contributes to decision making.
Galileo Galilei	1564-1642, Italian, astronomer, defensor of Copernicus' heliocentric theory, laid the foundations of the rational scientific method, condemned by the Inquisition.
Gaṇeśa	elephant-headed, son of Śiva and Pārvatī.
Foudre de Guerre	French expression applied to great war generals; Napoleon Buonaparte is described as a *foudre de guerre*.
Furore	Italian noun, indicates irresistible and passionate urge to do something, bad or good, evil or right; same root as *furia* or fury in English.
Gāyatrī	a poetic metre; *gāyatrī mantra* a particularly sacred *mantra, ṚV.*,

	III; 62, 10, *ṛṣi* Viśvāmitra.
Gestalt Psychology, Gestalt Theory	gestaltism stated briefly is the theory that ascertains that the outside world is first perceived and reacted to as a whole before any analysis or breakdown of the situation is done.
Ghee	butter oil, butter clarified by boiling.
Gṛaha praveśa	the rite of instalment of the home fire.
Guru	preceptor, spiritual leader; the disciple or *śiṣya* remained for years in the *guru*'s *āśrama* or hemitage for training.
Hermeneutics	the science of the "mysteries" or occult knowledge and of the art to decipher them, specially the ancient Egyptian Mysteries that aimed at inspiring wisdom in man as a pathway to eternity.
Hermes Trismegistus	thrice great Hermes; legendary individual, clan or God of ancient Egypt and keeper of traditional mysteries; the aspect of the Greek God Hermes, related to the ancient God Thot, representative of Re, the Supreme God.
Hominidae	the name of the family in the zoological nomenclature to which the single genus *Homo* belongs, the only species of the genus is *Homo sapiens* that is the scientific name of the present day human species.
Homo sapiens	scientific name of the human

	species; *Homo sapiens sapiens* is the scientific name of the sub-species to which we belong.
Hotṛ	chief priest who makes the invocations.
Ilā	food, libation of milk or refreshment, personified and sanctified; according to Sāyaṇa, the goddess presiding on the Earth; "*Ilā*'s sacred place" is the sacrificial fire.
Imra	partly identified with Indra by the Kalash people; God of creation; divider of the annual course of the sun, important personality of the winter solstice rites.
Indo-European	a varied assemblage of people whose languages have much in common, a similar syntax and probably common roots; the term was first used in the modern sense by R.G. Latham in 1851, with no reference to racial or national characteristics.
Indra	king of the Gods; probably a deified historical chief and leader.
Inoculum	Latin noun indicates a small portion of a pure microbial or tissue culture that is transferred to another culture medium to replicate and multiply the cultures.
Intelligĕre	Latin verb, to understand, to conceptualize, to visualize the connection between different concepts.

Jātavedas	all knowing, omniscient; an epithet of Lord Agni.
Jñānam	knowledge.
Kāfir	word used to designate non-muslim people, infidels.
Kalash	people of Kafiristan, now Nuristan, Gilgit, Hunza, most probably descendants of early migrant Āryan tribes.
Kalpa	a day and night of Brahmā, equal to 4,320,000,000 human years, also the ceremonial of the *Veda*.
Kāma	desire of good in general, power that satisfies desire, love.
Karma	principle of universal causality, cosmic law that is eternally recurrent and retributes the good and bad of the doer; the sum total of man's thoughts, feelings, desires and actions, the right and evil consequences of it may be effected in the present or in future incarnations.
Kawa	species of the genus *Hydnocarpus*, family Flacourtiaceae; oil extracted from the seeds has several medicinal properties, specific for leprosy and parasiticide.
Khorassan	a geographical area of present day Afghanistan.
Kirāta	non Āryan people, may designate one or more than one specific group of non-Āryan people, specially forest and mountain dwellers; in the *Rāmāyana*, they

	are described as islanders and coastal fish-eaters.
Knowledge	spiritual and intellectual understanding, has several aspects: *vidya*, learning, science; *prajña*, spiritual insight, wisdom; *sanjña*, comprehension, intellectual grasp, awareness; *jñāna*, contemplative knowledge. Opposite is *avidyā* or lack of knowledge and understanding.
Kuśika	the *ṛṣi* who founded the Kauśika family of Kanyākubja, from whom sage Viśwāmitra descended.
Lactase	enzime present in intestinal juices, hydrolizes lactose into dextrose and galactose.
Ladakh	north-eastern part of Jammu and Kashmir state of India having Tibetan culture.
Lāmā	Tibetan monks followers of *Mahāyāna* Buddhism.
Lavoisier, Antoine Laurent	1743-1794, French chemist, established the law of conservation of matter, executed under the French revolution.
Leonardo da Vinci	1452-1519, Italian artist, scientist, engineer, inventor, an extra- ordinarily gifted man.
Limonite	iron ore, hydrated ferrous oxide with about 30-32% iron.
Lohār	ironsmiths, traditionally nomadic.
Lopāmudrā	the wife of *mahārṣi* Agastya.
Maghavan	an epithet of Indra, Lord of Bounty
Mahā-bhūta	the five basic elements, earth,

	water, air, fire and *ākāśa*, ether or void.
Mahandeo	important God of the Kalash people, the name corresponds to Mahādeva, protector of the people, he safeguards, promises and protects, is messenger between men and the Creator; presiding deity of an important open air sacred area or temple that displays four prominent wooden sculptured horse heads, at Rumbur, just west of the Chitral area.
Malachite	copper ore, $CuCo_3, Cu(OH)_2$, dark green with differently coloured layers; called azurite when bluish.
Mantra	metrical poetic verses, sacred verses from the texts; main corpus of the *Vedic Saṁhitā*, most ancient and sacred part of the *Vedic* tradition.
Maruts	wind and storm gods, allies of Indra.
Menhir	prehistoric megaliths usually standing erect whose function may have been ceremonial and also as markers of astronomical phenomena.
Michelangelo Bonuarroti	1474-1563, Florence, Italian sculptor, architect, painter, poet, engineer.
Mitra	God of laws, contracts and correct behaviour, great enemy of falsehood, has a parallel in the *Avestan* God Mithra.

Mokṣa	salvation, final liberation.
Monism	negation of duality, essential identity of matter and spirit on mind.
Mudrā	seal, gesture; part of the position taken by the body to express particular ideas, words, emotions, sentiments.
Mythology	collection of myths and legends of all cultures, the science and study of myths.
Nāga	the cobra, the multiheaded cobra; the *Nāga* people.
Narāśaṁsa	the praise of men, an epithet of Lord Agni.
Nirukta	One of the *Vedāṅgas*, literal meaning is glossary or etymology; the text intends to explain and clarify the meaning of *Vedic* words; the *Nirukta* of Yāska mentions seventeen earlier writings, parts of the *Nirukta* may have been lost to us; the first critical study and translation was done by Roth whom R.N. Dandekar defines as the "grandsire of *Vedic* studies".
Ovum	the egg-cell, the mature female gamete or germ cell of plants and animals, the sperm cell or spermatozooid is the mature male gamete.
Pāda	a semi hemistich or line; each stanza or verse contains three or four *pādas*.
Pancagavya	the five gifts of the cow: dung,

	urine, milk, *ghee* and curd.
Pañcajanya	Kṛṣṇa's conch made from the shell of a sea demon called Pañcajana.
Pañcatantra	five books; famous Sanksrit story book in five parts; contains the teachings of Viṣṇuśarma to royal princes mainly through animal stories.
Paneer	coagulated milk casein, same as Italian *ricotta*.
Paṇinī	6th-5th century BC, Sanskrit grammarian whose masterpiece is the Aṣṭādhyāyī that fixed the rules of Sanskrit.
Paṇis	savage niggards, false, senseless, non worshipping, evil speaking thieves of cattle.
Parī	fairies.
Pariṇāma	theory of evolution in the *Sāṁkhya* system; in *Āyurveda* the *pariṇāmana* aspects of Agni are those of conversion of matter in the process of generation of heat of the food consumed and achievement of the benefits of nutrition.
Parjanya	the rain cloud, rain personified, rain God; also one of the Ādityas.
Pavamāna	self-purifying, an epithet of Lord Agni, see also Soma Pavamāna.
Pīpal	the sacred fig-tree or *aśvattha*, horse stand; the scientific name is *Ficus religiosa*; the Gods assemble under its shade and Viṣṇu was born under one,

associated to healing powers of nature and spiritual understanding, the bodh tree under which the Buddha received enlightenment.

Pitṛ — ancestors, Latin *patres*, fathers, manes, ancestral spirits; the *yajñas* to the *Pitṛs*, the *pitṛ-yajña* is one of the most sacred sacrifices to be performed at fixed periods and times.

pH — symbol to express the hydrogen-ion concentration (H^+) or hydroxyl-ion (OH^-) concentraion of a solution; indicates acidity or alcalinity of a solution.

placenta — the temporary organ or structure that attaches the embryo or fetus of mammals to the mother's uterus, similar structures in other animals and the tissues that attach the ovule to the ovary of the fruits of plants.

Polyphemus — one of the giant Cyclops of Greek mythology having only one eye in the middle of the forehead; Polyphemus was outwitted by Odysseus (Ulysses) and his companions whom he kept emprisoned in his cave; he was intoxicated with newly fermented grapes wine, blinded with the tip of a fire-hardened tree-trunk and the prisoners escaped hidden under the belly of Polyphemus' sheep that were endowed with

	very long wool.
Prajāpati	king of creation, title of some of the earliest gods; later elevated to the rank of a Supreme Creator God lord of Gods and Men.
Prākṛti	Nature, natural languages; the vernacular languages and dialects that emerged from the fusion of the *Ādibhāsa* or original language with the local ones; also provincial dialects of Sanksrit, some like Pāli retained root forms of ancient *Vedic* Sanskrit.
Pramantha	the drill or upper *araṇi* made with the wood from the *aśwattha* or sacred fig-tree, the *pīpal* or *Ficus religiosa*; Prometheus may have been conceptually the Greek counterpart of the *Pramantha*.
Prāṇa	breath, breath of life, wind.
Primum mobile	Latin expression meaning "the first cause".
Pṛthvi	the broad Earth, mother of all beings; our Earth is *bhūmi* also called *urvī*, wide.
Prometheus	the legendary man of Greek mythology who stole fire from the Gods to give it to man; he was condemned to be chained to the Caucasus mountains while a vulture unendingly feeds on his liver, may be the Greek counterpart of the *Pramantha*.
Puruṣa	original man; the principle that animates *Prakṛti*; the *puruṣa-sūkta* was the original *yajña* for

	the creation of the castes.
Pūṣan	the guide, the path-finder, the nourisher; protector of travellers and of cattle; ancient deity frequently associated with Indra and Bhaga, often counted among the Ādityas; guardian of living animals, able in recovering stolen goods and identifying robbers, prominent figure in the myth of the stolen cows hidden away in a cave by the Paṇis.
Pyrite	Iron ore, FeS_2
Ṛhbus	skillful, clever, industrious, three brothers divinized because of their wisdom, the eldest is Ṛbhukśan; descendants of Aṅgiras, expert artisans, specially of metal objects.
Ṛgveda	the first and ancientmost of the four *Veda*, has affinities with the *Avesta*.
Ṛṣi	ancient sages and poets, authors of *Vedic* hymns, the seven *ṛṣi* or *saptarṣi* are mind-born sons of Brahmā and are represented in the sky by the seven stars of the Great Bear; the *Śatapatha Brāhmaṇa* names them: Gautama, Bharadvāja, Viśvāmitra, Jamadagni, Vasiṣṭha, Kāśyapa, Atri; Bhṛgu is added to the seven in the *Vāyu Purāṇa*; the *Viṣṇu Purāṇa* adds Dakṣa and makes nine *Brahmā-ṛṣis*; more were later added to the original seven.

Ṛta	true, righteous, order and law of the Universe, the Cosmic Law and Order and Cosmic Ethics.
Rudra	the one who roars and howls, who is terrible; God of storms, father of Rudras and Maruts, both destructive and beneficial as healer of men and cattle; first called *Mahā-deva* in the *Yajurveda*; in the *Bṛhad-āraṇyaka-Upaniṣad* Rudras are the ten vital *prāṇa*; later identified with Śiva.
Rūmī	Mawlānā Jalāl al-Dīn, 1207-1273, born in Balkh, Khorassan; mystic, supreme *Sufi* poet of the persian language.
Sacred	from Latin sacrum, consecrated, holy, God esteemed, hallowed by specific rites, cannot be desecrated, dedicated to a God or deity.
Sacred forests	forests where a presiding deity resides; nothing can be removed from the sacred forests, they serve the purpose of nature reserves; the concept in varying forms can be found in most cultures where total or partial conservation of selected areas is enforced.
Sacrifacĕre	Latin verb, to make sacred, to sanctify by appropriate rites.
Sacrifice	rites and rituals, acts of giving offerings to Gods and Deities, sacrifices may involve ritual slaughtering of animals; first

	institutors of sacrifice were Aṅgiras, Atharvan, Bhṛgu, Manu.
Sacrum	Latin, meaning sacred as *sanctum* means saint and *sanctum sanctorum* means the sacred-most place of all the sacred spaces of a temple, sacred enclosure or sacred area.
Sādhu	ascetic who has acquired super-natural powers, may have acquired extraordinary powers through the practice of *yoga*, a *yogi*.
Sage	wise man, used as a synonym of *ṛṣi*.
Sahastrā-muṣkha	having a thousand testicles, epithet of Lord Indra.
Sakti	energy, power, the consorts of Gods and Deities who are his manifest power; the God and his consort; specifically the consort of Lord Śiva or Devī who is expressed in many forms.
Saṅgha	monastic order; association of equally minded persons; society.
Sāmaveda	the melodious *Veda*; the *Sāma-veda Saṁhitā* contains material mostly from the *ṚV*.
Saṁhitā	collection of *Vedic* hymns.
Śami	wood of the *śami* tree from which the lower *araṇi* is made, for this reason Agni is said to have the *śami* for its womb; the *araṇis* are also objects of worship.
Śaṅkara	788-820, Kaladi, Kerala; also called Śankarācārya; philo-

	sopher, greatest exponent of *Advaita*, a form of extreme monism.
Sannyāsī	those who have renounced the world; fourth *āśrama* of the *brāhmaṇa*'s religious life; religious mendicants and wanderers in general.
Sapēre	Latin verb, to know, to taste.
Sapientia	Latin for wisdom, knowledge, the gist of knowing; derived forms in English are: sapience for wisdom and knowledge; sapid for savory or tasty, opposite of tasteless or insipid, meaning with no salt.
Saptasindhu	land of the seven rivers: Gaṅgā, Yamunā, Saraswatī, Sutudrī, Paruṣṇī (now Ravi), Marud-vṛihā, Ārjīkīya or Vipāśa (now Beas); the land of the seven rivers, sapta-sindhava.
Saraswatī	flowing waters, a river and a Deity, the sacred river Saraswatī marked the early boundary of Āryāvarta and the personified Goddess was worshipped for bestowing fertility, wealth and health thanks to the cleansing power of its waters; the name later became that of the Goddess of learning, arts and sciences but the river dried out due to a curse, however remains of a "fossil" river may be seen on remote sensing imagery.
Sasarparī	running swiftly everywhere, may be an epithet of Vāk and daughter

	of Sūrya, as such she is intelligence that thinks rapidly; *Sasarparī* was gifted by the Jamadagnis because being intelligence she could dispel famine.
Śāstra	treatises, books having great authority, sacred texts, includes *śruti* and *smṛti*.
Savitṛ	vivifier, stimulator, generator: personification of Life and Hope, of the Sun as giver of life on Earth; a very ancient God and at times considered to be one of the Ādityas.
Sāyaṇa	prime minister of the king of Vijayanagar in the fourteenth century; the great scholiast who wrote a most important commentary of the *Veda* and who, according to H.H. Wilson "must have been in possession of all the interpretations which had been perpetuated by tradition".
Science	English from the Latin *scientia*; word given to methodically formulated knowledge acquired systematically.
Scīre	Latin verb, to know.
Seers	sages, the *ṛṣis*, they received the *śruti* or revealed Truth through visions.
Seven rivers	*saptasindhava*, see *Saptasindhu*.
Shaman	medecine man, magician, possessed of supernatural abilities; the man or woman

	endowed with magical powers that enable him or her to establish contact with super-natural entities of behalf of and to obtain benefits for human beings who do not have such powers.
Śiṣya	the disciple of a *guru*, bound to give unremitting and unquestionable obedience and service to his *guru* from whom he receives the highest teachings, training, guidance and knowledge.
Sītā	furrow, symbol of femininity, generosity, purity and good agricultural practices to whom one owes loayalty and faithfulness; daughter of Mother Earth, brought up by king Janaka and returned to Mother Earth to attest to her purity; personification of agricultural and vegetation cycles and fertility, Rāma's consort, mother of twins, Lava and Kuśa.
Śiva	the auspicious one, gradually emerged as *Mahādeva* and *Mahāyogī* though he is referred to only once and indirectly in the *ṚV*.
Śloka	stanza having a full meaning in itself.
Smṛti	what was been learned, remembered and transmitted by tradition; the *Vedāṅgas*, *Sūtra*, *Rāmāyaṇa*, *Mahābhārata*, *Purāṇas*, etc.
Solstice	the time and points in the ecliptic at which the sun is farthest away, north or south of the equator.

Species — Latin noun, same in English, used in biological sciences to designate all the individuals who are morphologically similar, who can reproduce sexually and whose descendants are fertile; chemical species are the chemical elements.

Stotra — hymns of prayers and praise of Gods, liturgical poems; the *Vedic* hymns are the greatest *stotras* of all.

Sudās — leader of the Bhārata, second wave of Āryan immigrants; great warrior and great statesman, he won over to his side the powerful Tṛtsus of whom he became king to jointly overpower the *Pañcajanāh* and the neighbouring non-Āryan principalities. The *purohita* of the Bhārata was Bharadvāja who was later replaced by Viśvāmitra while Vasiṣṭha was the *purohita* of the Tṛtsus, as a result of the alliance of the Bhāratas with the Tṛtsus, Vasiṣṭha replaced Viśvamitra as the *purohita* of the Bhāratas and the latter in great anger passed over to the Pañcajanāh. This was the origin of the great war of the Ten kings with the Bhāratas and Tṛtsus led by Sudās and Vasiṣṭha on one side and the "ten Kings" and Viśvāmitra on the other; described in the *ṚV* as the *Dāśarajña*.

Sufi — word of doubtful origin, already in

	use about 900-1000 AD, may be derived from the Arabic word *soof* that means wool and refers to the robe of early Muslim mystics; the word sufism was first used in 1821 by German scholars; the term *Sufi* is normally applied to enlightened saints whose spiritual experiences are much above the understanding of ordinary or learned men.
Sutudrī	former Satadru, flowing in hundred channels, one of the seven rivers of Saptasindhu, present day Sutlej, the Zaradrus of the Greek and Hesudrus of the Romans.
Śruti	revelation, what was heard, sub-tones of the Indian musical scale, revealed Truth, *Veda*, *Brāhmaṇa* and *Upaniṣad*.
Śūnya	zero, emptiness, nothingness.
Sūrya	sun deified and personalized, giver of light, warmth, life; one of the most referred to deities in the *Veda*, specially the *ṚV*; sun worship is common to most ancient cultures.
Svastika	meaning it is well, derived from *su* and *asti* where *ka* is a case-ending; the *svastika* may be derived from the *cakra*, wheel, or form the *araṇis* that are usually kept in a cross position; the right hand or clockwise *svastika* is considered to be masculine and auspicious, the left hand or

	counter clockwise; is considered to be feminine and unauspicious; common very ancient symbol throughout the world.
Tantra	different derivations proposed, very ancient teachings probably pre- *Vedic* and of obscure origin; many *tantric* texts are available but more have been kept secret although Tantrism contains some of the loftiest philosophies as well as unusual bodily practices.
Tanūnapāt	son of himself, an epithet of Lord Agni, fire can only be ignited or produced by fire.
Tapas	heat, ardour; active means for the acquisition of merit and spiritual power; improper translation: ascetism.
Thom, Alexander	1894-1985, Scotland; professor of Engineering Science, University of Oxford, 1945-1961; active life-long interest in prehistoric megalithic structures and sites of Britain and Brittany.
Thugs	secret fraternity of assassins, included both hindus and muslims.
Trita	very old *Vedic* God whose character is not clear, Trita Āptya appears to be related to atmospheric waters, brings rains; the word Āptya is etymologically connected with āp-water, has something to do with the preparation of *soma*; Thrita is known in the *Avesta* and is older

	than Indra, his functions appear to have been that to overcome the cloud demon, an achievement later imitated by Indra "in the manner of Trita" (*ṚV*.I; 52,5). Trita Āptya and Agni *Apām-napāt* are both born in waters, a feature shared by no other deity..
Tvaṣṭṛ	artisan par excellence, originally regarded as having fashioned the universe, in later mythology his functions were taken over by other Gods.
Ulysses	Latin name of the ancient Greek hero *Odysseus*, wisest and most cunning of the Greek heros of the Trojan war.
Ur-Gott	German word, original God.
Vāk-Vāc	Vāc, speech, word, Speech personified.
Vaisvānara	beloved of all, dear to all men, belonging to all men, an epithet of Lord Agni.
Vālakhilya	eleven hymns of *ṚV Maṇḍala* VIII, not included in Sāyaṇa's commentaries, usually interpolated after No. 48 and numbered 49 59.
Vanaspati	Lord of the forest, tall tree, sovereign tree, forest, plants.
Varadā	bestower of boons, an epithet of Saraswatī.
Varuṇa	one who encompasses everything, name derived from root *vṛ* meaning "to bind": Varuṇa bound and regulated the universe because he possessed *asu*, the

	universal magic potence, the *Asura* par excellence, the Saṁrāt who enforced and upheld Ṛta, the cosmic law.
Vasiṣṭha	most wealthy; one of the seven *Ṛṣis*, seer of many hymns of the *ṚV*, belonged to the Tṛtsu tribe, celebrated in the *Rm*, the *MBh*, the *P*; all the hymns of *maṇḍala* VII are by Vasiṣṭha.
Vāyu	air, wind; often associated with Indra, resides in the air; said to have been originated by the breath of *Puruṣa*; king of *gandharvas*.
Vyāsa	Veda Vyāsa, compiler, arranger, reputed to have arranged the *Vedas* and compiled the *Mahābhārata*.
Vidyā	knowledge, wisdom, discrimination, understanding.
Vipās	unbound, present day river Beas, former Byas, Hyphasis of the Greek; one of the seven rivers; the Vipās refused to drown sage Vasiṣṭha who had thrown himself bound up into the river as he wished to die; the river returned him unbound-*vipāsa*, on its banks.
Viśvacārṣaṇih	God of all men, an epithet of Lord Agni.
Visvedevah	all Gods, preservers of men and bestowers of rewards; many hymns are addressed to them; they later constitute a special class of ten Gods: Vasu, Satya,

	Kratu, Dakṣa, Kāla, Kāma, Dhṛti, Kuru, Purūravas and Mādravas.
Viśvāmitra	friend of all, one of the seven *Ṛṣi,* born a *Kṣatrya* he elevated himself to brahamanhood through extraordinary *tapas;* author of most hymns of *Maṇḍala* III of the *ṚV*; mentioned both as Kuśika descendent of Kusa or of Gādhi; has a prominent position in the *Rm* and the *MBh*, life long enemy of sage Vasiṣṭha, the final reconciliation of the two arch-enemies is given in the *Rm* (see Vasiṣṭha-Sudās)
Vratapati	Lord ov vows, an epithet of Lord Agni.
Yajamāna	the institutor of a *yajña* or sacrifice.
Yajurveda	the fourth *Veda,* burrows extensively from the *ṚV*. may be a later addition to the earlier three *Veda.*
Yakṣa	meaning uncertain
Yajña	*Vedic* sacrifice, personified in the *Purāṇas* as son of Ruchi and husband of Dakṣiṇā, has the head of a deer and was killed at Dakṣa's sacrifice.
Yajus	sacrificial text and formula.
Yoga	one of six schools of philosophy, founded by sage Yājñavalkya who is also the author of the *White YV*, the *SPBr*, the *Br-Up* and other *Smṛti.*
Yūpa	sacrificial post, usually made of

daru or *deodar* wood, with a knob at the top for fixing a wheel on which the priest or the *yajamāna* sat and were whirled around three times on the occasion of a very ancient rite, the *dūrcana* meaning difficult climb, symbolizing the elevation from the profane to the sacred; the rite was associated also with mounting the chariot, hence the title of *cakravartin* given to the great ancient emperors; the *dwaja* or flagstaff erected in front of temples may be a later development of the *yūpa*.

Bibliography and References

Andrewartha, H.G. and L.C. Birch, 1954, *The Distribution and abundance of animals*, XV-782 pp., Univ. of Chicago Press, Chicago.

Anonymous, 1987, Iroquois, in: *National Geographic Magazine*, 172 (3), 370-473.

Arranos, *circa* 95-175 AD, *Storia di Alessandro*, ed. comm.. by L. Belloni, 1980, 471pp. 1 map, Rusconi, Milano.

Brain, C.K. and A. Sillen, 1988, Evidence from the Swartkrans cave for the earliest use of fire, *Nature*, 336; 464-466.

Carson, R. 1962, *Silent Spring*, Houghton Mifflin, U.S.A.

Dandekar, R.N. 1979a, *Vedic Mythological Tracts, Select Writings I*, 383 pp., Ajanta Publ., Delhi at Bhandarkar Inst. Press, Poona.

Dandekar, R.N. 1979b, *Insights into Hinduism, Select Writing II*, 394 pp., Ajanta Publ., Delhi at Bhandarkar Inst. Press, Poona.

Dandekar, R.N. 1981, *Excercises in Indology, Select Writings III*, 385 pp., Ajanta Publ., Delhi at Bhandarkar Inst. Press, Poona.

Dash, Vd. Bhagwan, 1971, *Concept of Agni in Āyurveda*, 212 pp., *Chowkhamba Sanskrit Studies*, v. 81, Varanasi.

Dash, Vd. Bhagwan, 1975, *Embryology and Maternity in Āyurveda*, 147 pp., Diary Publ., Delhi.

Ferguson, L. 1988, A Catalogue of the Alexander Thom Archive held in the National Monuments Record of Scotland, pp. 31-131, in: *Records in Stone*, ed. by C.L.N. Ruggles, 1988, XVII+519 pp., Cambridge University Press.

Filliozat, J. 1955, L'apport de l'étude des religions de l'Inde à la science de l'homme, in: *NUMEN II. Suppl.*, Leiden, pp. 108-119.

Filliozat, J. 1991, *Religion, Philosophy, Yoga-Collected articles*. Motilal Banarsidass, Varanasi, 484 pp.

Bergaigne, A. 1878-1883, *La religion védique d'après les hymnes du Rig-Veda*, Vols. I-III.

Gadgill, M. and M.D. Subash Chandran, 1992, Sacred Groves, 183-187, in: *Indigenous Vision*, 304 pp., India International Centre and Sage Publications, New Delhi.

Ganapati, S.V. 1982, *Sāmaveda*, 491 pp., Motilal Banarsidass.

Gause, G.F. 1934, *The Struggle for Existence*, Williams and Williams, Baltimore 163 pp.

Gause, G.F. 1935, Vérifications expérimentales de la théorie mathématique de la lutte pour la vie, in: *Actualit'ees scientifiques et industrielles*,277, 61 pp.

Griffith, R.T.H. 1889, *The Humns of the Ṛgveda*,707 pp. 1986, Motilal Banarsidass, Delhi.

Griffith, R.T.H. 1893, repr. 1991, *The Sāmaveda Saṁhitā*, 500 pp., Nag Publishers, New Delhi.

Griffith, R.T.H. 1899, repr. 1990, *Yajurveda Saṁhitā*, 620 pp., Nag Publishers, New Delhi.

Hale, W.E. 1991, *Asura in early Vedic Religion.*, 226 pp., Motilal Banarsidass, Varanasi.

Hillebrandt, A. 1876, *Über die Göttin Aditi*, Bresslau.

Hillebrandt., A. 1877, *Varuna und Mitra*, Bresslau.

Hillebrandt, A. 1891, *Vedische Mythologie I. Soma und verwandte Götter*, Bresslau, English translation S.R.

Sarma: *Vedic Mythology*, 1980, vol. 1, 476 pp; 1981, v. 2, 488 pp., Motilal Banarsidass, Delhi,

Hutchinson G.E. 1959, Il concetto moderno di nicchia ecologica, in: *Memorie Istituto Italiano di Idrobiologia*, 11.

Hutchinson, G.E. 1961, The Paradox of the Plankton, *American Naturalist*, 95; 137-145.

Jain, S.K. 1985, *Medicinal Plants of India*.Book Trust of India, 180pp.

Keith, A.B. 1920, repr. 1981, *Ṛgveda Brāhmanas; the Aitreya and Kauṣītaki Brāhmanas of the Ṛgveda*, 555 pp. Motilal Banarsidass.

Kohler, W. 1947, *Gestalt Psychology*, Liveright Publishing Corporation, 222pp., New York.

Lièvre, V. and J-Y Loude, 1990, *Le Chamanisme des Kalash du Pakistan.*, ed. CNRS, Paris, 558 pp.

Loude, J.-Y and V. Lièvre, 1984, *Solstice païen*, Presses de la Renaissance. Paris, 361 pp.

Loude, J.-Y and V. Lièvre, 1987, Fêtes d'été chez les Kalash du Nord Pakistan, in: *L'éthnographie*, 100-01, pp. 191-220.

Ludwig, A. 1876-88, *Der Ṛgveda*. 6 vols. Prag.

Mackie, E. 1988, Investigating the Prehistoric Solar Calendar, pp. 206-231, in: *Records in Stone*, ed. by C.L.N. Ruggles, 1988, XVII+519 pp., Cambridge University Press.

Mandler, J.M. 1990, A new Perspective in Cognitive Development in Infancy, in: *American Scientist*, 78, 3, 236-243.

Margalef, D.R. 1956, La teoria de la información en ecologia, in: *Memorias Real Academia de Ciencias y Artes.*, Barcelona, 32; 373-449.

Margalef, D.R. 1958a, Temporal succession and spatial heterogeneity in plankton, in: *Perspectives in Marine Biology*, A. A. Buzzati-Traverso (ed.) Univ. of

California, Berkeley.

Margalef, D.R. 1958b, *Information Theory in Ecology*,2nd ed., W.B. Saunders Co., Philadelphia.

Marazzi, U. 1984, *Testi dello sciamanesimo siberiano e centro-asiatico*, ed. UTET-TEA, Torino, 610 pp.

Müller, M. 1849-1874, *Ṛg-Veda Saṁhitā, the Sacred Hymns of the Brāhmanas, together with the commentary of Sayanacārya.* 6 vols. 2nd ed. London,

Müller, F.M. and H. Oldenberg, 1895, *Vedic Hymns*, No. 32 and 46 of the Series: *Sacred Books of the East* ed. by F.M. Müller 1979 ed., repr. 1990, Atlantic Publ. and Distr., New Delhi, v.1, XV+476 pp; v. 2, X+500 pp.

Nadkarni, K.M. 1976 ed. *Indian Materia Medica*, v.1, 1319 pp; v.2, 968 pp., Popular Prakashan Pvt. Ltd., Bombay.

Odum, E.P. 1971, Fundamentals of Ecology, 3rd ed., W.B. Saunders Co. Philadelphia, XIV+574 pp.

Patil, D.R. 1946, repr. 1973, *Cultural History from the Vāyu Purāṇa*, 348 pp. Motilal Banarsidass, Varanasi.

Piaget, J. 1952, *The Origin of Intelligence in Children*, transl. M. Cook, International University Press.

Radin, P. 1941, *La Religion Primitive, sa nature et son origine*, Gallimard, Paris, 242 pp.

Ramakrishnan, P.S. 1992, *Shifting Agriculture and Sustainable Development*, XVII+424 pp., *Man and Biosphere Series*, v. 10, UNESCO, Paris.

Renou, L. 1956, *Hymnes spéculatifs du Veda*, 276 pp., UNESCO, Ser: *Connaissance de l'Orient*, Collection UNESCO d'Oeuvres Représentatives. Gallimard, Paris.

Robertson, G.S. 1888, *The Kafirs of the Hindukush*, Facsimile repr., by Sang e-Meel Publ., Lahore, 656 pp. 1 map.

Ruggles, C.L.N. 1988 (ed.) *Records in Stone. Papers in memory of Alexander Thom*, XVII+519 pp., Cambridge University Press.

Saint-Exupéry, A. Le Petit Prince. 1921.

Sastry, T.S.K. 1985, *Vedānga Jyotiṣa of Lagadha; in its Ṛk and Yajus Recensions*, 74 pp., Indian National Science Academy, Delhi,

Schuré E. 1899, *Les Grands Initiés.*, ed. Laterza, 1989, I Grandi Iniziati, 480 pp.

Sharma, B.N.K. 1978, *The Brahmasūtras and their principal Commentaries*, 3 vols., Bharatiya Vidya Bhavan, Bombay.

Sivananda, Swami, 1977, 2nd ed. *Brahma Sūtra*, 686 pp., Motilal Banarsidass.

Vannucci, M. 1989, *The Mangroves and Us.*, 203 pp., Indian Association for the Advancement of Science, New Delhi.

Vannucci, M. 1991, Sacred Groves or Holy Forests. No. 32, in: *Concepts of Space, Ancient and Modern*, ed. by Kapila Vatsyayan, Indira Gandhi National Centre for the Arts, 665 pp.

Vannucci, M. 1992, Tradition and Change, pp. 23-34, in: *Indigenous Vision*, 304 pp. ed. Geeti Sen, Sage Publ. Inc., India and India International Centre, New Delhi.

Vannucci, M. (in press) Fire and Marksmanship as transcultural values, in: Seminar on Transcultural Values, organized by Kapila Vatsyayan for the Indira Gandhi National Centre for the Arts, New Delhi.

Vannucci M. (in press), Conscious and Unconscious Time. A Biological Perspective, in: International Seminar on *Kāl*, Time, organized by Kapila Vatsyayan for the Indira Gandhi National Centre for the Arts, New Delhi.

Vannucci M. and H.S. Bisht, (in preparation) On two Megaliths from Kumaon.

Vatsyayan, K. 1983, *The Square and the Circle of the Indian Arts*, XI+159 pp., Roli Books International, New Delhi.

Vesci, U.M. 1985, *Heat and Sacrifice in the Vedas*, 335 pp., Motilal Banarsidass, Varanasi.

Wanninayaka, P.B. 1982, *Āyurveda in Sri Lanka*, 88 pp. Ministry of Health Publications, Colombo.

Whitney, W.D. 1905, 1984 repr. *The Atharva-Veda Saṁhitā*, 2 v., 1046 pp., Motilal Banarsidass, Varanasi.

Volterra, V. 1926, Variazioni e fluttuazioni del numero di individui in speci animali conviventi, in: *Memorie Reale Accademia dei Lincei*, 6; 2; 181-187.

Volterra, V. 1928, Variations and fluctuations of the number of individuals in animal species living together, in: *Journal du Conseil*, 3; 1-51.

Index

Errata

Page	Para-graph	Line	Printed	To be read as
22	3	8	widsom	wisdom
25	1	3	your	you
28	1	2	mathe- matician	mathematician
29	1	1	nd	and
	2	10	ominiscience	omniscience
30	3	20	this studies	these studies
31	2	12	ignorance	ignorance.
33	3	17	is that order	is that of order
37	2	17	does that	do that
48	2	1	God, primordial	God, the primordial
52	2	4	grassland firs	grassland fires
56	1	2	everything originally	everything, originally
56	3	8	is a conti nuous	is a continuous
66	2	4	of the Altai	of the Altaï
68	2	4	acquired science)	acquired science)"
69	2	20	bones domestic	bones of domestic
70	1	14	No only	Not only
71	1	20	um is	drum is
71	2	5	medicinal ṕlants	medicinal plants
72	2	6	unfúrled	unfurled
86	1	6	Many authros	Many authors
90	3	9	verý ancient	very ancient
92	1	15	is prayed equally	is prayed to equally
93	2	23	density grow and	density grows and
96	3	1	*Sahastramuṣka*	*Sahaśramuṣka*
	3	6	*Sahastramanyo*	*Sahaśramanyo*

99	3	3	interfers	interferes
102	3	6	water is substance	water is the substance
105	1	22	he five ecological	the five ecological
106	1	4	idefinitely	indefinitely
	2	1	symapthetic magic	sympathetic magic
118	4	4	during episode	during episodes
121	2	3	of object that	of objects that
124	2	l.l.	then the looked	then he looked
125	1	6	(*ṚV.* III; 26, 7.	(*ṚV.* III; 26, 7).
136	1	l.l.	by by the ancient	by the ancient
	2	9	"*ācāryas* who came	*ācāryas* who came
145	2	10	better good,	better food,
146	3	6/7	b ycommentators	by commentators
147	2	5	Agni Vaiśhvānara	Agni Vaiśvānara
148	2	1	to the Schuré	to Schuré
157	3	5	his magnificient	his magnificent
161	4	3	he conrtrols	he controls
162	3	11/12	110, 3)	110, 3)
163	1	1	thetruthful.	the truthful.
166	1	5	thesc relationship	these relationships
167	2	10	be aware to	be aware of
185		Michelangelo Bonuarroti		Michelangelo Buonarroti
186	Monism	2	and spirit on mind	and spirit or mind
192			*Sahastrāmuṣkha*	*Sahaśramuṣkha*
195	Sītā	4	loayalty	loyalty
197	*Svastika*	5	form	from
199	Ulysses	3	Greek heros	Greek heroes
209	s.c.	l.l.	iron melting	iron ore smelting
210			Agoresavara	Agoresvara
			Benodebi,	Banodebi,
211			Dadhikarās	Dadhikrās
	s.c.	l.l.	Francis	St. Francis
212	God		existance	existence
213	Man		origion	origin
214			Rama Krishnar	Ramakrishnan, P.S.
216	Vāk		faminine	feminine
	Yoni		faminity	femininity

l. l. = last line s. c. = second column